MEMOIRS
of the
American Mathematical Society

Number 1023

n-Harmonic Mappings between Annuli

The Art of Integrating Free Lagrangians

Tadeusz Iwaniec
Jani Onninen

July 2012 • Volume 218 • Number 1023 (first of 5 numbers) • ISSN 0065-9266

American Mathematical Society
Providence, Rhode Island

Library of Congress Cataloging-in-Publication Data

Iwaniec, Tadeusz.

n-harmonic mappings between annuli: The art of integrating free Lagrangians / Tadeusz Iwaniec, Jani Onninen.

p. cm. — (Memoirs of the American Mathematical Society, ISSN 0065-9266 ; no. 1023)

"July 2012, volume 218, number 1023 (first of 5 numbers)."

Includes bibliographical references.

ISBN 978-0-8218-5357-3 (alk. paper)

1. Quasiconformal mappings. 2. Extremal problems (Mathematics). I. Onninen, Jani, 1973- II. Title.

QA360.I936 2011

516.3′62—dc23 2012007086

Memoirs of the American Mathematical Society

This journal is devoted entirely to research in pure and applied mathematics.

Publisher Item Identifier. The Publisher Item Identifier (PII) appears as a footnote on the Abstract page of each article. This alphanumeric string of characters uniquely identifies each article and can be used for future cataloguing, searching, and electronic retrieval.

Subscription information. Beginning with the January 2010 issue, *Memoirs* is accessible from `www.ams.org/journals`. The 2012 subscription begins with volume 215 and consists of six mailings, each containing one or more numbers. Subscription prices are as follows: for paper delivery, US\$772 list, US\$617.60 institutional member; for electronic delivery, US\$679 list, US\$543.20 institutional member. Upon request, subscribers to paper delivery of this journal are also entitled to receive electronic delivery. If ordering the paper version, subscribers outside the United States and India must pay a postage surcharge of US\$69; subscribers in India must pay a postage surcharge of US\$95. Expedited delivery to destinations in North America US\$61; elsewhere US\$167. Subscription renewals are subject to late fees. See `www.ams.org/help-faq` for more journal subscription information. Each number may be ordered separately; *please specify number* when ordering an individual number.

Back number information. For back issues see `www.ams.org/bookstore`.

Subscriptions and orders should be addressed to the American Mathematical Society, P. O. Box 845904, Boston, MA 02284-5904 USA. *All orders must be accompanied by payment.* Other correspondence should be addressed to 201 Charles Street, Providence, RI 02904-2294 USA.

Memoirs of the American Mathematical Society (ISSN 0065-9266) is published bimonthly (each volume consisting usually of more than one number) by the American Mathematical Society at 201 Charles Street, Providence, RI 02904-2294 USA. Periodicals postage paid at Providence, RI. Postmaster: Send address changes to Memoirs, American Mathematical Society, 201 Charles Street, Providence, RI 02904-2294 USA.

This publication is indexed in *Science Citation Index*®, *SciSearch*®, *Research Alert*®, *CompuMath Citation Index*®, *Current Contents*®/*Physical, Chemical & Earth Sciences.*
Printed in the United States of America.

∞ The paper used in this book is acid-free and falls within the guidelines established to ensure permanence and durability.
Visit the AMS home page at `http://www.ams.org/`

10 9 8 7 6 5 4 3 2 1 17 16 15 14 13 12

Contents

Abstract

The central theme of this paper is the variational analysis of homeomorphisms $h\colon \mathbb{X} \xrightarrow{\text{onto}} \mathbb{Y}$ between two given domains $\mathbb{X}, \mathbb{Y} \subset \mathbb{R}^n$. We look for the extremal mappings in the Sobolev space $\mathscr{W}^{1,n}(\mathbb{X}, \mathbb{Y})$ which minimize the energy integral

$$\mathscr{E}_h = \int_{\mathbb{X}} \| Dh(x) \|^n \, \mathrm{d}x.$$

Because of the natural connections with quasiconformal mappings this n-harmonic alternative to the classical Dirichlet integral (for planar domains) has drawn the attention of researchers in Geometric Function Theory. Explicit analysis is made here for a pair of concentric spherical annuli where many unexpected phenomena about minimal n-harmonic mappings are observed. The underlying integration of nonlinear differential forms, called *free Lagrangians*, becomes truly a work of art.

Received by the editor December 7, 2010.

Article electronically published on September 19, 2011; S 0065-9266(2011)00640-4.

2000 *Mathematics Subject Classification.* Primary 30C65, 30C75, 35J20.

Key words and phrases. n-Harmonics, Extremal problems, Quasiconformal mappings, Variational integrals.

Iwaniec was supported by the National Science Foundation grant DMS-0800416 and the Academy of Finland project 1128331, and Onninen by the National Science Foundation grant DMS-1001620. A part of this research was done when the first author was visiting the University of Michigan, as Gehring visiting professor. He thanks the Department of the University of Michigan for the support and hospitality.

Preface

The future developments in modern geometric analysis and its governing partial differential equations (PDEs) will continue to rely on physical and geometric intuition. In recent years, this trend has become more pronounced and has led to increasing efforts of pure and applied mathematicians, engineers and other scientists, to share the ideas and problems of compelling interest. The present paper takes on concrete questions about energy minimal deformations of annuli in $\mathbb{R}^n$. We adopted the interpretations and central ideas of nonlinear elasticity where the applied aspects of our results originated. A novelty of our approach is that we allow the mappings to slip freely along the boundaries of the domains. It is precisely in this setting that one faces a real challenge in establishing the existence, uniqueness and invertibility properties of the extremal mappings. The underlying concept of *Free Lagrangians* is the core of the matter.

Our approach is purely mathematical though the questions are intimately derived from Nonlinear Elasticity. Both the theoretical and practical aspects of this work culminate in actual construction of the mappings with smallest conformal energy. Special efforts have been devoted to somewhat subtle computational details to present them as simply and clearly as possible.

We believe the final conclusions shed considerable new light on the Calculus of Variations, especially for deformations that are free on the boundary. We also feel that some new facts discovered here have the potential for applications in Geometric Function Theory as well as for better understanding the mathematical models of Nonlinear Elasticity.

CHAPTER 1

Introduction and Overview

1. Basic notation

Let us take a moment to recall a very much needed notation from the calculus of vector fields and matrix fields. Consider a mapping $h\colon \mathbb{X} \overset{\text{into}}{\longrightarrow} \mathbb{Y}$ between domains $\mathbb{X} \subset \mathbb{R}^n$ and $\mathbb{Y} \subset \mathbb{R}^m$, $h = (h^1, h^2, \dots, h^m)$, where $h^1, \dots, h^m$ are scalar functions in the Sobolev space $\mathscr{W}^{1,p}_{\text{loc}}(\mathbb{X})$. The differential $Dh(x)$, defined at almost every $x \in \mathbb{X}$, represents a linear transformation of $\mathbb{R}^n$ into $\mathbb{R}^m$, $Dh(x)\colon \mathbb{R}^n \to \mathbb{R}^m$. With the standard choice of the coordinates in $\mathbb{R}^n$ and $\mathbb{R}^m$ we have a matrix field, again denoted by Dh,

$$Dh = \begin{bmatrix} h^1_{x_1} & h^1_{x_2} & \cdots & h^1_{x_n} \\ \vdots & \cdots & \vdots & \\ h^m_{x_1} & h^m_{x_2} & \cdots & h^m_{x_n} \end{bmatrix} \in \mathscr{L}^p_{\text{loc}}(\mathbb{X}, \mathbb{R}^{m\times n}). \tag{1.1}$$

Hereafter, we abbreviate the notation of distributional partial derivative $\frac{\partial F}{\partial x_i}$ to F_{x_i}, $i = 1, \dots, n$ for $F \in \mathscr{L}^1_{\text{loc}}(\mathbb{X})$. The differential matrix, also called Jacobian matrix or *deformation gradient*, acts on a vector field $V = (V^1, V^2, \dots, V^n) \in \mathscr{L}^q_{\text{loc}}(\mathbb{X}, \mathbb{R}^n)$ by the rule

$$[Dh]V = \begin{bmatrix} h^1_{x_1} & h^1_{x_2} & \cdots & h^1_{x_n} \\ \vdots & \cdots & \vdots & \\ h^m_{x_1} & h^m_{x_2} & \cdots & h^m_{x_n} \end{bmatrix} \begin{bmatrix} V^1 \\ \vdots \\ V^n \end{bmatrix} = \begin{bmatrix} \langle \nabla h^1, V\rangle \\ \vdots \\ \langle \nabla h^m, V\rangle \end{bmatrix} \in \mathscr{L}^1_{\text{loc}}(\mathbb{X}, \mathbb{R}^m).$$

Here ∇ stands for the gradient operator acting on real-valued functions in $\mathscr{W}^{1,p}_{\text{loc}}(\mathbb{X})$. More generally, consider an arbitrary matrix field

$$M = \begin{bmatrix} M^1_1(x) & \cdots & M^1_n(x) \\ \vdots & \cdots & \vdots \\ M^m_1(x) & \cdots & M^m_n(x) \end{bmatrix} \in \mathscr{L}^p_{\text{loc}}(\mathbb{X}, \mathbb{R}^{m\times n}) \tag{1.2}$$

and denote its row-vector fields by $\mathfrak{r}^1, \dots, \mathfrak{r}^m \in \mathscr{L}^p_{\text{loc}}(\mathbb{X}, \mathbb{R}^n)$. Similarly, the column-vector fields will be denoted by $\mathfrak{c}_1, \dots, \mathfrak{c}_n \in \mathscr{L}^p_{\text{loc}}(\mathbb{X}, \mathbb{R}^m)$. The divergence operator acting on a vector field $\mathfrak{r} = (r_1, \dots, r_n) \in \mathscr{L}^p_{\text{loc}}(\mathbb{X}, \mathbb{R}^n)$ is a Schwartz distribution defined by

$$\operatorname{div} \mathfrak{r} = \sum_{i=1}^n \frac{\partial r_i}{\partial x_i} \in \mathcal{D}'(\mathbb{X}, \mathbb{R}).$$

Then the divergence of a matrix field $M \in \mathscr{L}^p_{\mathrm{loc}}(\mathbb{X}, \mathbb{R}^{m\times n})$ is a distribution, valued in $\mathbb{R}^m$,

$$\mathrm{Div} M = \begin{bmatrix} \mathrm{div}\, \mathfrak{r}^1 \\ \vdots \\ \mathrm{div}\, \mathfrak{r}^m \end{bmatrix} \in \mathcal{D}'(\mathbb{X}, \mathbb{R}^m).$$

In particular,

$$\mathrm{Div} Dh = \Delta h \in \mathcal{D}'(\mathbb{X}, \mathbb{R}^m)$$

where $\Delta = \frac{\partial^2}{\partial x_1^2} + \cdots + \frac{\partial^2}{\partial x_n^2}$ is the usual Laplacian. The matrix fields $M \in \mathscr{L}^p_{\mathrm{loc}}(\mathbb{X}, \mathbb{R}^{m\times n})$ which satisfy the equation $\mathrm{Div} M \equiv 0$ will be called *divergence free*, meaning that

$$\int_{\mathbb{X}} \langle M, D\eta \rangle = 0 \quad \text{for every test mapping } \eta \in \mathscr{C}^\infty_\circ(\mathbb{X}, \mathbb{R}^m).$$

Hereafter $\langle A, B \rangle = \mathrm{Tr}(A^* B)$ is the inner product of matrices. We will be typically working with the Hilbert-Schmidt norm of a matrix

$$\| M \|^2 = \langle M, M \rangle = \sum_{j=1}^m \sum_{i=1}^n |M_i^j|^2.$$

2. Mathematical model of hyperelasticity

Geometric Function Theory (GFT) is currently a field of enormous activity where the language and general framework of Nonlinear Elasticity is very helpful. As this interplay develops, the n-harmonic deformations become well acknowledged as a possible generalization of mappings of finite distortion. We have also found a place for n-harmonic deformations in the theory of nonlinear hyperelasticity. J. Ball's fundamental paper [**5**] accounts for the principles of this theory and sets up mathematical models. Historically, the relation between hyperelasticity and quasiconformal theory, has not been clearly manifested, but it is indeed very basic and fruitful.

One can roughly describe the hyperelasticity as a study of weakly differentiable homeomorphisms $h \colon \mathbb{X} \xrightarrow{\text{onto}} \mathbb{Y}$ between domains in $\mathbb{R}^n$ (or n-manifolds) that minimize a given energy integral,

$$\mathscr{E}_h = \int_{\mathbb{X}} \mathbf{E}(x, h, Dh)\, \mathrm{d}x < \infty \qquad Dh \colon \mathbb{X} \to \mathbb{R}^{n\times n}, \tag{1.3}$$

The condition on the injectivity of h is imposed in order to avoid interpenetration of matter. The Jacobian matrix $Dh(x) \in \mathbb{R}^{n\times n}$, defined at almost every point $x \in \mathbb{X}$, is referred to as the *deformation gradient.* In this model the so-called *stored energy function* $\mathbf{E} \colon \mathbb{X} \times \mathbb{Y} \times \mathbb{R}^{n\times n} \to \mathbb{R}$ is given; it characterizes mechanical properties of the elastic material in $\mathbb{X}$ and $\mathbb{Y}$.

Motivated by GFT we will be essentially concerned with the n-harmonic energy, also called *conformal energy*

$$\mathscr{E}_h = \int_{\mathbb{X}} \| Dh(x) \|^n\, \mathrm{d}x. \tag{1.4}$$

Another energy integral of interest in GFT is

$$\mathscr{F}_h = \int_{\mathbb{X}} \frac{\| Dh(x) \|^n}{|h(x)|^n}\, \mathrm{d}x. \tag{1.5}$$

We devote Chapter 8 to this latter integral.

3. Variational integrals in GFT

In another direction, we recall Geometric Function Theory in $\mathbb{R}^n$ and its governing variational integrals. Let as begin with a conformal mapping $h : \mathbb{X} \xrightarrow{\text{onto}} \mathbb{Y}$. Thus at every $x \in \mathbb{X}$ we have the relation between the norm of the Jacobian matrix and its determinant $\| Dh(x) \|^n = n^{\frac{n}{2}} J(x,h)$. This can be expressed in the form of a *nonlinear Cauchy-Riemann system* of PDEs;

$$D^*h \cdot Dh = J(x,h)^{\frac{2}{n}} \, \mathbf{I} \tag{1.6}$$

It is evident that the n -harmonic energy of $h : \mathbb{X} \xrightarrow{\text{onto}} \mathbb{Y}$ depends only on the deformed configuration. Indeed, we have

$$\mathscr{E}_h = \int_{\mathbb{X}} \| Dh(x) \|^n \, \mathrm{d}x \;=\; n^{\frac{n}{2}} \int_{\mathbb{X}} J(x,h) \, \mathrm{d}x \;=\; n^{\frac{n}{2}} |\mathbb{Y}| \tag{1.7}$$

For other homeomorphisms $g : \mathbb{X} \xrightarrow{\text{onto}} \mathbb{Y}$, in the Sobolev space $\mathscr{W}^{1,n}(\mathbb{X}, \mathbb{Y})$, we only have a lower bound, due to Hadamard's inequality for determinants:

$$\mathscr{E}_g = \int_{\mathbb{X}} \| Dg(x) \|^n \, \mathrm{d}x \geqslant n^{\frac{n}{2}} \int_{\mathbb{X}} J(x,g) \, \mathrm{d}x = n^{\frac{n}{2}} |\mathbb{Y}|$$

Thus conformal deformations $h : \mathbb{X} \xrightarrow{\text{onto}} \mathbb{Y}$ are none other than those having the n-harmonic energy equal to $n^{\frac{n}{2}}|\mathbb{Y}|$, the smallest possible. It is for this reason that conformal mappings are frequently characterized as *absolute minimizers* of the n-harmonic integral. However, it is rare in higher dimensions that two topologically equivalent domains are conformally equivalent, because of Liouville's rigidity theorem. Even in the plane, multiply connected domains like annuli are of various conformal type. From this point of concerns Quasiconformal Theory [**3, 29**] offers significantly larger class of mappings.

DEFINITION 1.1. A homeomorphism $h : \mathbb{X} \to \mathbb{R}^n$ of Sobolev space $\mathscr{W}^{1,1}_{\text{loc}}(\mathbb{X}, \mathbb{R}^n)$ is said to have *finite outer distortion* if

$$\| Dh(x) \|^n \leqslant n^{\frac{n}{2}} K(x) \, J(x,h)$$

for some measurable function $1 \leqslant K(x) < \infty$. The smallest such $K(x)$ is called the outer distortion, denoted by $\mathbb{K}_{_O}(x,h)$. Then h is K *-quasiconformal* if $\mathbb{K}_{_O}(x,h) \leqslant K$ for some constant K.

A concept somewhat dual to outer distortion is the inner distortion. For this, consider the cofactor matrix $D^\sharp h$ (it represents infinitesimal deformations of $(n-1)$-dimensional entities) defined for invertible Dh via Cramer's rule

$$D^\sharp h \cdot D^*h = J(x,h) \, \mathbf{I}$$

For a map $h \in \mathscr{W}^{1,1}_{\text{loc}}(\mathbb{X}, \mathbb{R}^n)$, not necessarily a homeomorphism, but with non-negative Jacobian, we introduce the *inner distortion function*

$$\mathbb{K}_{_I}(x,h) = \begin{cases} \dfrac{\|D^\sharp h(x)\|^n}{n^{n/2} \, [J(x,h)\,]^{n-1}} & \text{if } \; J(x,h) > 0 \\ 1 & \text{if } \; J(x,h) = 0 \end{cases} \tag{1.8}$$

REMARK 1.2. Any map of finite outer distortion has finite inner distortion and $\mathbb{K}_{_I}(x) \leqslant \mathbb{K}_{_O}(x)^{n-1}$, but not vice versa.

It is again interesting to find the place for such mappings in continuum mechanics. The latter deals with the positive definite matrix $\mathbf{C}(x) = D^*h(x)\,Dh(x)$ as the *right Cauchy-Green* deformation tensor. While on the other hand, there is a fundamental interplay between mappings of finite distortion and the *Beltrami equation*

$$D^*h(x)\,Dh(x) \;=\; J(x,h)^{\frac{2}{n}}\;\mathbf{G}(x)\,, \qquad \det \mathbf{G}(x) \equiv 1 \tag{1.9}$$

Thus $\mathbf{G} = \mathbf{G}(x)$, called the *distortion tensor* of h, is none other than the Cauchy-Green tensor renormalized so as to have determinant identically equal to one. The symmetric positive definite matrix function $\mathbf{G} = \mathbf{G}(x) = [G_{ij}(x)] \in \mathbb{R}^{n\times n}$ can be viewed as a Riemann metric tensor on $\mathbb{X}$. In this way h becomes conformal with respect to this, usually only measurable, metric structure on $\mathbb{X}$. Thus $\mathbf{G}(x)$ is uniformly elliptic in case of K-quasiconformal mappings.

It is in this Riemannian manifold framework that variational interpretations of quasiconformal mappings really crystalize. For example, the solutions to the Beltrami equation (1.9) are none other then the *absolute minimizers* of their own energy integrals. Indeed, a homeomorphism $h : \mathbb{X} \to \mathbb{Y}$ of Sobolev class $\mathscr{W}^{1,n}(\mathbb{X},\mathbb{Y})$ solves the Beltrami equation (1.9) if and only if

$$\mathscr{E}_h \overset{\text{def}}{=\!=} \int_{\mathbb{X}} \mathbf{E}(x, Dh)\,\mathrm{d}x \;=\; n^{\frac{n}{2}} \int_{\mathbb{X}} J(x,h)\,\mathrm{d}x \;=\; n^{\frac{n}{2}}\,|\mathbb{Y}| \tag{1.10}$$

where the integrand is defined on $\mathbb{X}\times\mathbb{R}^{n\times n}$ by the rule

$$\mathbf{E}(x,\xi) = \;\big(\operatorname{Tr}\,[\,\xi\;\mathbf{G}^{-1}(x)\;\xi^*\,]\big)^{\frac{n}{2}}\;, \qquad \xi \in \mathbb{R}^{n\times n}$$

As in the conformal case, for all other homeomorphisms $g\colon \mathbb{X} \xrightarrow{\text{onto}} \mathbb{Y}$, in the Sobolev space $\mathscr{W}^{1,n}(\mathbb{X},\mathbb{Y})$, we have the lower bound $\mathscr{E}_g \geqslant n^{\frac{n}{2}}[\mathbb{Y}]$. The most appealing conclusion is a connection between the n -harmonic energy of $h : \mathbb{X} \to \mathbb{Y}$ and the inner distortion function of the inverse mapping $f = h^{-1} : \mathbb{Y} \to \mathbb{X}$.

PROPOSITION 1.3. (TRANSITION TO THE INVERSE MAP)
Let $f \in \mathscr{W}^{1,n-1}_{\text{loc}}(\mathbb{Y},\mathbb{X})$ be a homeomorphism of finite outer distortion between bounded domains, with $\mathbb{K}_{_I}(y,f) \in \mathscr{L}^1(\mathbb{Y})$. Then the inverse map $h = f^{-1} : \mathbb{X} \xrightarrow{\text{onto}} \mathbb{Y}$ belongs to the Sobolev class $\mathscr{W}^{1,n}(\mathbb{X},\mathbb{Y})$ and we have

$$n^{\frac{n}{2}} \int_{\mathbb{Y}} \mathbb{K}_{_I}(y,f)\,dy \;=\; \int_{\mathbb{X}} \|Dh(x)\|^n\,dx \tag{1.11}$$

This identity gains in significance if we realize that the polyconvex variational integrand in the left hand side turns into a convex one, a rarity that one can exploit when studying quasiconformal mappings of smallest $\mathscr{L}^1$ -mean distortion. From yet another perspective, it is worth mentioning the classical Teichmüller theory which is concerned, broadly speaking, with extremal mappings between Riemann surfaces. The extremal Teichmüller mappings are exactly the ones whose distortion function has the smallest possible $\mathscr{L}^\infty$-norm. The existence and uniqueness of such an extremal quasiconformal map within a given homotopy class of quasiconformal mappings is the heart of Teichmüller's theory. Now, in view of the identity (1.11), minimizing the $\mathscr{L}^1$-norm of the inner distortion function offers a study of n-harmonic mappings. Is there any better motivation?

4. Conformal energy

For $\mathbb{X}$ and $\mathbb{Y}$ open regions in $\mathbb{R}^n$, we shall consider mappings

$$h = \left(h^1, ..., h^n\right) : \mathbb{X} \longrightarrow \mathbb{Y} \tag{1.12}$$

in the Sobolev class $\mathscr{W}^{1,n}(\mathbb{X}, \mathbb{Y})$. Thus the Jacobian matrix of h and its determinant are well defined at almost every point $x \in \mathbb{X}$. We recall the notation

$$Dh = \left[\frac{\partial h^i}{\partial x_j}\right] \in \mathscr{L}^n(\mathbb{X}, \mathbb{R}^{n\times n}), \qquad J(x,h) = \det Dh \in \mathscr{L}^1(\mathbb{X}) \tag{1.13}$$

Here, as usual, $\mathbb{R}^{n\times n}$ is supplied with the inner product and the Hilbert-Schmidt norm:

$$\langle A, B\rangle = \operatorname{Tr}\,(A^\star B) = \sum_{ij=1}^{n} A^i_j B^i_j \qquad \| A \| = \langle A, A\rangle^{\frac{1}{2}} \tag{1.14}$$

At the initial stage of our undertaking the *n-harmonic integral* will be subjected to the orientation preserving homeomorphisms $h : \mathbb{X} \xrightarrow{\text{onto}} \mathbb{Y}$, so that

$$\mathscr{E}_h = \int_{\mathbb{X}} \| Dh(x) \|^n \, dx \geqslant n^{\frac{n}{2}} \int_{\mathbb{X}} J(x,h)\, dx = n^{\frac{n}{2}} |\mathbb{Y}| \tag{1.15}$$

In dimension $n \geqslant 3$, it may well be that no homeomorphism $h : \mathbb{X} \xrightarrow{\text{onto}} \mathbb{Y}$ of finite n-harmonic energy exists, as the following result [**31**] shows.

THEOREM 1.4. *Let $\mathbb{X} \subset \mathbb{R}^n$ be a ball with a k-dimensional closed disk removed, and let $\mathbb{Y} \subset \mathbb{R}^n$ be a ball with a $(k+1)$-dimensional closed disk removed, $1 \leqslant k < n-1$. Then every homeomorphism $h : \mathbb{X} \xrightarrow{\text{onto}} \mathbb{Y}$ has infinite n-harmonic energy.*

Note that both $\mathbb{X}$ and $\mathbb{Y}$ are *topological annuli*; that is, homeomorphic images of a spherical annulus $\mathbb{A} = \{x \colon r < |x| < R\}$. Let us view the disks removed from the balls as cracks. It can be easily shown, by means of an example, that mappings of finite conformal energy may outstretch an $(n-1)$-dimensional crack into an n-dimensional hole. However, Theorem 1.4 ensures us that, in principle, mappings of finite energy cannot increase the dimension of lower dimensional cracks, a fact highly nontrivial to observe and prove [**31**]. From now on we assume without explicit mention that the domains $\mathbb{X}$ and $\mathbb{Y}$ admit at least one homeomorphism $h : \mathbb{X} \xrightarrow{\text{onto}} \mathbb{Y}$ in the Sobolev space $\mathscr{W}^{1,n}(\mathbb{X}, \mathbb{Y})$

5. Weak limits of homeomorphisms

But the true challenge is to find a deformation $h^\circ : \mathbb{X} \xrightarrow{\text{onto}} \mathbb{Y}$ with the smallest possible energy. In general, when passing to the weak limit of the minimizing sequence of homeomorphisms, the injectivity of the extremal map will be lost. Nevertheless, from the point of view of the elasticity theory [**1, 5, 10**], such limits are still legitimate deformations to consider. For, if this is the case, they create no new cracks or holes in $\mathbb{Y}$. Let $\mathscr{P}(\mathbb{X}, \mathbb{Y})$ denote the class of weak limits of homeomorphisms $h : \mathbb{X} \xrightarrow{\text{onto}} \mathbb{Y}$ in the Sobolev space $\mathscr{W}^{1,n}(\mathbb{X}, \mathbb{Y})$. We refer to such limits as *permissible deformations.*[1] Differentiability and geometric features of permissible mappings are not as clear as one may have expected. In Theorems 1.5, 1.6 and 1.7

[1]Homeomorphisms converging weakly in $\mathscr{W}^{1,n}(\mathbb{X}, \mathbb{Y})$ also converge c-uniformly, so their limits are still continuous, taking $\mathbb{X}$ into $\overline{\mathbb{Y}}$.

we assume that $\mathbb{X}$ and $\mathbb{Y}$ are bounded domains of the same topological type, like spherical annuli, having at least two though finitely many boundary components. We consider a sequence $h_j : \mathbb{X} \xrightarrow{\text{onto}} \mathbb{Y}$ of homeomorphisms converging weakly in $\mathscr{W}^{1,n}(\mathbb{X}, \mathbb{Y})$ to a mapping $h : \mathbb{X} \to \overline{\mathbb{Y}}$. In general, homeomorphisms $h_j : \mathbb{X} \xrightarrow{\text{onto}} \mathbb{Y}$ do not extend as continuous maps to the closure of $\mathbb{X}$, but the distance functions $x \to \operatorname{dist}\big(h_j(x), \partial\mathbb{Y}\big)$ do extend. This is also true for the limit mapping h. The precise result to which we are referring is the following:

THEOREM 1.5. [**30**, Theorem 1.1] *For the above-mentioned pair of domains $\mathbb{X}$ and $\mathbb{Y}$, there exists a nonnegative continuous function $\eta = \eta(x)$ defined on $\overline{\mathbb{X}}$ such that*

$$\operatorname{dist}\big(h_j(x), \partial\mathbb{Y}\big) \leqslant \eta(x) \, \| Dh_j \|_{\mathscr{L}^n(\mathbb{X})}, \qquad \eta \equiv 0 \text{ on } \partial\mathbb{X}. \tag{1.16}$$

In particular,

$$\operatorname{dist}\big(h(x), \partial\mathbb{Y}\big) \leqslant \eta(x) \sup_{j \geqslant 1} \| Dh_j \|_{\mathscr{L}^n(\mathbb{X})}. \tag{1.17}$$

The weak limit h actually covers the target domain, but this may fail if $\mathbb{X}$ and $\mathbb{Y}$ have only one boundary component.

THEOREM 1.6. [**30**, Theorem 1.4] *The mapping h is continuous and $\mathbb{Y} \subset h(\mathbb{X}) \subset \overline{\mathbb{Y}}$. Furthermore, there exists a measurable mapping $f : \mathbb{Y} \to \mathbb{X}$, such that*

$$h \circ f = \mathrm{id} : \mathbb{Y} \to \mathbb{Y},$$

everywhere on $\mathbb{Y}$. This right inverse mapping has bounded variation,

$$\| f \|_{\mathrm{BV}(\mathbb{Y})} \leqslant \int_{\mathbb{X}} \| Dh(x) \|^{n-1} \, \mathrm{d}x.$$

As noted in [**30**, Remark 9.1] the weak limit h is monotone in the sense of C.B. Morrey [**37**].

THEOREM 1.7. *The mapping h is monotone, meaning that for every continuum $\mathbb{K} \subset \mathbb{Y}$ its preimage $h^{-1}(\mathbb{K}) \subset \mathbb{X}$ is also a continuum; that is compact and connected.*

The proof of this theorem is presented in Section 3.

6. Annuli

The first nontrivial case is that of doubly connected domains. Thus we consider mappings $h : \mathbb{A} \to \mathbb{A}^*$ between concentric spherical annuli in $\mathbb{R}^n$.

$$\mathbb{A} = \mathbb{A}(r, R) = \{x \in \mathbb{R}^n \; ; \quad r < |x| < R\} \;, \qquad 0 \leqslant r < R < \infty$$
$$\mathbb{A}^* = \mathbb{A}(r_*, R_*) = \{y \in \mathbb{R}^n \; ; \quad r_* < |y| < R_*\} \;, \quad 0 \leqslant r_* < R_* < \infty$$

Such domains are of different conformal type unless the ratio of the two radii is the same for both annuli. As for the domains of higher connectivity in dimension $n = 2$, the conformal type of a domain of connectivity $\ell > 2$ is determined by $3\ell - 6$ parameters, called Riemann moduli of the domain.[2] This means that two ℓ-connected domains are conformally equivalent if and only if they agree in all $3\ell - 6$ moduli. But we shall have considerably more freedom in deforming $\mathbb{X}$ onto $\mathbb{Y}$, simply by means of mappings of finite energy. An obvious question to ask

[2]In this context the mappings are orientation preserving.

is whether minimization of the n-harmonic integral is possible within homeomorphisms between domains of different conformal type.

Concerning uniqueness, we note that the energy $\mathscr{E}_h$ is invariant under conformal change of the variable $x \in \mathbb{A}$. Such a change of variable is realized by *conformal automorphism* of the form

$$x' = \left(\frac{rR}{|x|^2}\right)^k Tx \tag{1.18}$$

where $k = 0, 1$ and T is an orthogonal matrix.

7. Hammering a part of an annulus into a circle, $n = 2$

Let us caution the reader that a minimizer $h^\circ \colon \mathbb{A} \to \mathbb{A}^*$, among all permissible deformations, does not necessarily satisfy the Laplace equation. A loss of harmonicity occurs exactly at the points where h° fails to be injective. This is the case when the target annulus $\mathbb{A}^*$ is too thin as compared with $\mathbb{A}$; precisely, if

$$\frac{R_*}{r_*} < \frac{1}{2}\left(\frac{R}{r} + \frac{r}{R}\right), \qquad \text{-}\textit{annuli below the Nitsche bound} \tag{1.19}$$

By way of illustration, consider the so-called *critical Nitsche map*

$$\aleph(z) = \frac{1}{2}\left(z + \frac{1}{\bar{z}}\right), \qquad 0 < |z| < \infty \tag{1.20}$$

This harmonic mapping takes an annulus $\mathbb{A}(1, R)$ univalently onto $\mathbb{A}^* = \mathbb{A}(1, R_*)$, where $R_* = \frac{1}{2}\left(R + \frac{1}{R}\right)$. We have equality at (1.19), and $\aleph$ is the energy minimizer. Note the symmetry $\aleph\left(\frac{1}{\bar{z}}\right) = \aleph(z)$. Thus the same Nitsche map takes reflected annulus $\mathbb{A}(R^{-1}, 1)$ univalently onto $\mathbb{A}^*$. Let us paste these two annuli along their common boundary

$$\mathbb{A} \overset{\text{def}}{=\!=} \mathbb{A}(r, R) = \mathbb{A}(r, 1] \cup \mathbb{A}[1, R), \qquad r = \frac{1}{R}$$

Now the same harmonic map $\aleph : \mathbb{A} \overset{\text{onto}}{\longrightarrow} \mathbb{A}\,[1, R_*)$ is a double cover. Its Jacobian determinant vanishes along the unit circle, the branch set of $\aleph$. Therefore, $\aleph$ is not permissible (it is not a weak $\mathscr{W}^{1,2}$-limit of homeomorphisms). An extension of $\aleph : \mathbb{A}(1, R) \overset{\text{onto}}{\longrightarrow} \mathbb{A}^*$ inside the unit disk to a permissible mapping of $\mathbb{A}(r, R)$ onto $\mathbb{A}^*$ can be nicely facilitated by squeezing $\mathbb{A}(r, 1)$ onto the unit circle. This procedure will hereafter be referred to as hammering the inner portion of the domain annulus onto the inner boundary of the target. Precisely, the map we are referring to takes the form

$$h^\circ(z) = H(|z|)\,\frac{z}{|z|} \overset{\text{def}}{=\!=} \begin{cases} \frac{z}{|z|} & \frac{1}{R} < |z| \leqslant 1\,, \qquad \text{hammering part} \\ \frac{1}{2}\left(z + \frac{1}{\bar{z}}\right) & 1 \leqslant |z| \leqslant R\,, \qquad \text{harmonic part} \end{cases}$$

It is true, though somewhat less obvious, that: h° is a $\mathscr{W}^{1,2}$-limit of homeomorphisms $h \colon \mathbb{A} \overset{\text{onto}}{\longrightarrow} \mathbb{A}^*$ and its energy is smaller than that of any homeomorphism from $\mathbb{A}$ onto $\mathbb{A}^*$, see Figure 1.1.

We shall actually prove the following theorem.

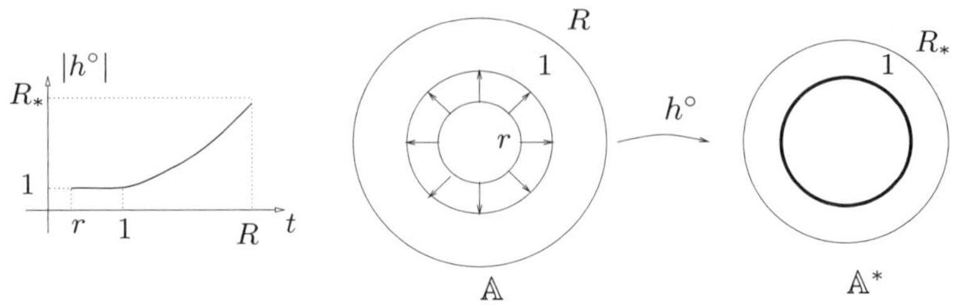

FIGURE 1.1. Hammering the inner ring into the unit circle.

THEOREM 1.8. *Let $\mathbb{A} = \mathbb{A}(r, R)$ and $\mathbb{A}^* = \mathbb{A}(r_*, R_*)$ be planar annuli, $0 < r < R < \infty$ and $0 < r_* < R_* < \infty$. We have:*
Case 1. (Within the Nitsche bound) If

$$\frac{R_*}{r_*} \geqslant \frac{1}{2}\left(\frac{R}{r} + \frac{r}{R}\right)$$

then the harmonic homeomorphism

$$h^\circ(z) = \frac{r_*}{2}\left(\frac{z}{r} + \frac{r}{\bar{z}}\right), \qquad h^\circ \colon \mathbb{A} \xrightarrow{\text{onto}} \mathbb{A}^*$$

attains the smallest energy among all homeomorphisms $h\colon \mathbb{A} \xrightarrow{\text{onto}} \mathbb{A}^*$, and as such is unique up to a conformal automorphism of $\mathbb{A}$.
Case 2. (Below the Nitsche bound) If

$$\frac{R_*}{r_*} < \frac{1}{2}\left(\frac{R}{r} + \frac{r}{R}\right)$$

then the infimum energy among all homeomorphisms $h\colon \mathbb{A} \xrightarrow{\text{onto}} \mathbb{A}^*$ is not attained. Let a radius $r < \sigma < R$ be determined by the equation

$$\frac{R_*}{r_*} = \frac{1}{2}\left(\frac{R}{\sigma} + \frac{\sigma}{R}\right) \qquad \text{-critical Nitsche configuration.}$$

Then the following mapping

$$h^\circ(z) = \begin{cases} r_* \frac{z}{|z|} & r < |z| \leqslant \sigma \\ \frac{r_*}{2}\left(\frac{z}{\sigma} + \frac{\sigma}{\bar{z}}\right) & \sigma \leqslant |z| < R \end{cases}$$

is a $\mathscr{W}^{1,2}$-limit of homeomorphisms $h_j\colon \mathbb{A} \xrightarrow{\text{onto}} \mathbb{A}^*$, and its energy is smaller than that of any homeomorphism $h\colon \mathbb{A} \xrightarrow{\text{onto}} \mathbb{A}^*$.

The proof of this theorem was first given in [**2**]; here in Section 10 we present another one, based on free Lagrangians.

Let us emphasize that this result does not rule out the existence of univalent harmonic mappings from $\mathbb{A}$ onto $\mathbb{A}^*$, simply because harmonic homeomorphisms need not be the ones that minimize the energy. Nonexistence of harmonic homeomorphisms between such annuli $\mathbb{A}$ and $\mathbb{A}^*$ was conjectured by J. C. C. Nitsche [**41**]. After several partial results were obtained in [**17, 34, 35, 50**], the conjecture was proved in [**25, 26**]. The connection between the Nitsche conjecture and minimal surfaces is further explored in [**28**]. Hammering also occurs in free boundary problems for minimal graphs, where it is called *edge-creeping* [**9, 21, 46**]. Similar hammering phenomena will be observed in higher dimensions as well.

8. Principal n-harmonics

In studying the extremal deformations between spherical annuli it is natural to look for the radially symmetric solutions of the n-harmonic equation

$$\operatorname{div}\left(\|Dh\|^{n-2} Dh\right) = 0\,, \qquad h(x) = H\big(|x|\big)\,\frac{x}{|x|} \tag{1.21}$$

There is a nice reduction of this problem to the first order (nonlinear) differential equation for the strain function $H = H(t)$, called the *characteristic equation* for H,

$$\mathcal{L}H = \left[H^2 + \frac{t^2 \dot{H}^2}{n-1} \right]^{\frac{n-2}{2}} \left(H^2 - t^2 \dot{H}^2 \right) \equiv \text{const.} \tag{1.22}$$

Although this equation provides a very convenient tool for studying properties of radial n-harmonics, a little caution is needed because $\mathscr{C}^1$-solutions to (1.22) may fail to satisfy the original equation (1.21). We shall distinguish four so-called *principal n-harmonics* $\aleph_\circ$, $\aleph^\circ$, $\aleph_+$ and $\aleph_-$. The first two are conformal mappings

$$\aleph_\circ(x) = x \quad \text{and} \quad \aleph^\circ(x) = \frac{x}{|x|^2} \tag{1.23}$$

The other two are more involved, but still radially symmetric

$$\aleph_+ = H_+\big(|x|\big)\frac{x}{|x|} \quad \text{and} \quad \aleph_- = H_-\big(|x|\big)\frac{x}{|x|} \tag{1.24}$$

They cannot be described in any elementary way, except for the case $n = 2$.

$$\aleph_\circ(z) = z\,,\ \aleph^\circ(z) = \frac{1}{\bar{z}} \tag{1.25}$$

$$\aleph_+(z) = \frac{1}{2}\left(z + \frac{1}{\bar{z}}\right) \quad \text{and} \quad \aleph_-(z) = \frac{1}{2}\left(z - \frac{1}{\bar{z}}\right) \tag{1.26}$$

The principal n-harmonics $\aleph_+$ and $\aleph_-$ are determined by solving the following Cauchy problems for their strain functions

$$\mathcal{L}H_+ \equiv 1 \qquad H_+(1) = 1 \tag{1.27}$$

$$\mathcal{L}H_- \equiv -1 \qquad H_-(1) = 0 \tag{1.28}$$

It has to be emphasized that all radial solutions to the n-Laplace equation take the form

$$h(x) = \lambda\,\aleph(kx)\,, \qquad \lambda \in \mathbb{R}\,,\ k > 0 \tag{1.29}$$

where $\aleph$ is one of the four principal solutions. We then see that every radial solution (originally defined in an annulus) extends n-harmonically to the entire punctured space $\mathbb{R}^n_\circ = \mathbb{R}^n \setminus \{0\}$, and is evidently $\mathscr{C}^\infty$-smooth. Their connections with (1.26) motivate our calling $\aleph_\circ$, $\aleph^\circ$, $\aleph_+$ and $\aleph_-$ the Nitsche maps in $\mathbb{R}^n$.

9. Elasticity of stretching

There is an important entity associated with the radial mappings; namely the elasticity of stretching

$$\eta_{_H}(t) = \frac{t\,\dot{H}(t)}{H(t)}\,, \qquad \text{provided } H^2 + \dot{H}^2 \neq 0 \tag{1.30}$$

The elasticity function of a power stretching $H(t) = t^\alpha$ is a constant, equal to α. Two particular cases $\eta \equiv 1$ and $\eta \equiv -1$ correspond to the conformal maps $\aleph_\circ$ and $\aleph^\circ$. In both cases $\mathcal{L}H \equiv 0$. There are exactly three types of radial n-harmonics:

(a) *Conformally Expanding.* These are the mappings with $|\eta_H| > 1$, everywhere. Equivalently, $\mathcal{L}H \equiv \text{const} > 0$. Geometrically it means that h exhibits greater change in the radial direction than in spherical directions.
(b) *Conformally Contracting.* These are the mappings with $|\eta_H| < 1$, everywhere. Equivalently, $\mathcal{L}H \equiv \text{const} < 0$.
(c) *Conformally Balanced.* These are conformal mappings with $|\eta_H| \equiv 1$, or, equivalently $\mathcal{L}H \equiv 0$.

For every permissible radial map $h(x) = H\big(|x|\big)\frac{x}{|x|} : \mathbb{A} \to \mathbb{A}^*$ (weak $\mathscr{W}^{1,n}$-limit of homeomorphisms) the elasticity function does not change sign. Indeed, this follows from the identity $J(x,h) = \dot{H}\big(|x|\big) \cdot \left(\frac{H(|x|)}{|x|}\right)^{n-1}$. The following $\mathscr{L}^1$-mean (with respect to the conformal density on $\mathbb{A}$) is equal to the ratio of the moduli of the annuli, see (1.40) for the definition of the modulus,

$$\fint_{\mathbb{A}} |\eta_H(x)| \, \mathrm{d}\mu(x) = \frac{\operatorname{Mod}\mathbb{A}^*}{\operatorname{Mod}\mathbb{A}}, \qquad \mathrm{d}\mu = \frac{\mathrm{d}x}{|x|^n} \quad [3] \tag{1.31}$$

In this way we are led to three types of pairs of the annuli $\mathbb{A}$ and $\mathbb{A}^*$.

(1) *Conformally expanding*; it pertains to a pair of annuli such that

$$\operatorname{Mod}\mathbb{A}^* > \operatorname{Mod}\mathbb{A}$$

(2) *Conformally contracting*; it pertains to a pair of annuli such that

$$\operatorname{Mod}\mathbb{A}^* < \operatorname{Mod}\mathbb{A}$$

(3) *Conformally equivalent*; these are the annuli having the same modulus

$$\operatorname{Mod}\mathbb{A}^* = \operatorname{Mod}\mathbb{A}$$

These three cases will be treated by using somewhat different estimates.

10. Conformally expanding pair

We shall show that the principal solution $\aleph_-(x) = H_-\big(|x|\big)\frac{x}{|x|}$ generates all minimizers of the conformal energy. It is rather easy to show that for a given expanding pair $\mathbb{A}$, $\mathbb{A}^*$ there exist unique $k > 0$ and $\lambda > 0$ such that the n-harmonic map $h^\circ(x) = \lambda\,\aleph_-(kx)$ takes $\mathbb{A}$ homeomorphically onto $\mathbb{A}^*$. However, the answer to the question whether this map minimizes the n-harmonic energy among all homeomorphisms is not obvious. When $n = 2$ or $n = 3$, the answer is "yes".

THEOREM 1.9. *Let* $\operatorname{Mod}\mathbb{A}^* > \operatorname{Mod}\mathbb{A}$. *Then for* $n = 2, 3$, *the n-harmonic radial map* $h^\circ = \lambda\aleph_-(kx)$ *assumes the minimum conformal energy within all homeomorphisms. Such a minimizer is unique up to a conformal automorphism of* $\mathbb{A}$.

Surprisingly, for $n \geqslant 4$ the answer will depend on how wide is the target annulus $\mathbb{A}^*$, relatively to $\mathbb{A}$.

---

[3] The integral mean notation $\fint_{\mathbb{E}} f(x)\,\mathrm{d}\mu(x)$ stands for the ratio $\int_{\mathbb{E}} f\,\mathrm{d}\mu / \int_{\mathbb{E}} \mathrm{d}\mu$.

THEOREM 1.10. *For dimensions $n \geqslant 4$, there exists a function $\mathcal{N}^\dagger = \mathcal{N}^\dagger(t)$, $t < \mathcal{N}^\dagger(t) < \infty$ for $t > 0$, see Figure 1.2, such that: if*

$$\operatorname{Mod}\mathbb{A} \leqslant \operatorname{Mod}\mathbb{A}^* \leqslant \mathcal{N}^\dagger(\operatorname{Mod}\mathbb{A}) \quad \text{-upper Nitsche bound for } n \geqslant 4, \tag{1.32}$$

then the map $h^\circ : \mathbb{A} \to \mathbb{A}^*$ is a unique (up to an automorphism of $\mathbb{A}$) minimizer of the conformal energy among all homeomorphisms.

THEOREM 1.11. *In dimensions $n \geqslant 4$, there are annuli $\mathbb{A}$ and $\mathbb{A}^*$ such that no radial stretching from $\mathbb{A}$ onto $\mathbb{A}^*$ minimizes the conformal energy.*

11. Conformally contracting pair

In this case we obtain the minimizers from the principal solution $\aleph_+ = H_+(|x|)\frac{x}{|x|}$. As before, we observe that the mappings

$$h^\circ(x) = \lambda\, H_+(kx)\,, \quad k > 0\,, \quad \lambda > 0 \tag{1.33}$$

are radial n-harmonics. Recall that $\mathbb{A}^*$ is conformally thinner than $\mathbb{A}$. But it is not enough. In contrast to the previous case, such n-harmonic mappings take the annulus $\mathbb{A}$ homeomorphically onto $\mathbb{A}^*$ only when $\mathbb{A}^*$ is not too thin. The precise necessary condition reads as

$$\text{the lower Nitsche bound;} \qquad \mathcal{N}_\dagger(\operatorname{Mod}\mathbb{A}) \leqslant \operatorname{Mod}\mathbb{A}^* \leqslant \operatorname{Mod}\mathbb{A}. \tag{1.34}$$

Numerically, the lower Nitsche function $\mathcal{N}_\dagger$ is given by

$$0 < \aleph_\dagger(t) = \omega_{n-1}\, \log H_+\left(\exp\frac{t}{\omega_{n-1}}\right) < t \qquad \text{for } \ 0 < t < \infty. \tag{1.35}$$

THEOREM 1.12. *Under the condition at (1.34) there exist unique $k > 0$ and $\lambda > 0$ such that $h^\circ(x) = \lambda H_+(kx)$ takes $\mathbb{A}$ homeomorphically onto $\mathbb{A}^*$. This map is a unique (up to a conformal automorphism of $\mathbb{A}$) minimizer of the conformal energy among all homeomorphisms of $\mathbb{A}$ onto $\mathbb{A}^*$.*

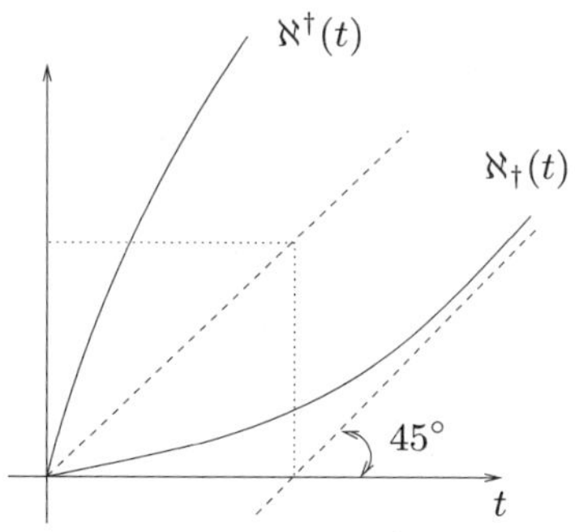

FIGURE 1.2. The Nitsche functions.

Rather unexpectedly, the injectivity of the weak limit of a minimizing sequence of homeomorphisms $h_j \colon \mathbb{A} \xrightarrow{\text{onto}} \mathbb{A}^*$ will fail once the lower bound in (1.34) is violated. Let us look more closely at the critical configuration of annuli; that is,

$$\mathcal{N}_\dagger(\operatorname{Mod}\mathbb{A}') = \operatorname{Mod}\mathbb{A}^*\,, \qquad \begin{cases} \text{where } \mathbb{A}' = \mathbb{A}(1,R) \text{ and} \\ \mathbb{A}^* = \mathbb{A}(1,R_*)\,, \quad R_* = H_+(R) \end{cases} \tag{1.36}$$

Thus the critical Nitsche map $\aleph_+(x) = H_+(|x|)\,\frac{x}{|x|}$, defined for $1 \leqslant |x| \leqslant R$ takes $\mathbb{A}'$ homeomorphically onto $\mathbb{A}^*$. The Jacobian determinant of $\aleph$ vanishes for $|x| = 1$. Now let us built a pair of annuli below the lower bound at (1.34), simply by pasting an additional spherical ring to $\mathbb{A}'$ along the unit sphere. Thus, we consider a slightly fatter annulus

$$\mathbb{A} = \mathbb{A}\,(r, R) = \mathbb{A}(r, 1] \cup \mathbb{A}[1, R)$$

As in dimension $n = 2$, we have

THEOREM 1.13. *The following deformation*

$$h^\circ(x) \stackrel{\text{def}}{=\!=} \begin{cases} \frac{x}{|x|} & r < |x| \leqslant 1\,, \qquad \textit{hammering part} \\ \aleph_+(x) & 1 \leqslant |x| \leqslant R\,, \qquad n\textit{-harmonic part} \end{cases} \tag{1.37}$$

is a $\mathscr{W}^{1,n}$-limit of homeomorphisms $h\colon \mathbb{A} \xrightarrow{\text{onto}} \mathbb{A}^*$ and its energy is smaller than that of any homeomorphism from $\mathbb{A}$ onto $\mathbb{A}^*$. Among all such mappings h° is unique up to conformal automorphisms of $\mathbb{A}$.

12. The conformal case $\operatorname{Mod}\mathbb{A} = \operatorname{Mod}\mathbb{A}^*$

Obviously the minimizers are conformal mappings. Precisely, they take the form

$$h^\circ(x) = \frac{\sqrt{r_* R_*}}{|x|} \left(\frac{\sqrt{rR}}{|x|}\right)^{\pm 1} Tx \tag{1.38}$$

where T is an orthogonal transformation. This case receives additional consideration in Section 15.

13. The energy function $\mathscr{F}_h$

So far we considered the domains $\mathbb{X}$ and $\mathbb{Y}$ equipped with the Euclidean metric. However, one may ask analogous questions for different metrics on $\mathbb{X}$ and $\mathbb{Y}$. We do not enter into a general framework here, but instead illustrate this possibility by introducing a conformal density on the target space, which we continue to assume to be a spherical annulus

$$\mathbb{Y} = \mathbb{A}^* = \mathbb{A}(r_*\,,\, R_*)\,, \quad \text{equipped with the measure } \mathrm{d}\mu(y) = \frac{\mathrm{d}y}{|y|^n} \tag{1.39}$$

The modulus of $\mathbb{A}^*$ is none other than its "conformal" volume

$$\operatorname{Mod}\mathbb{A}^* = \int_{\mathbb{A}^*} \mathrm{d}\mu(y) = \int_{\mathbb{A}^*} \frac{\mathrm{d}y}{|y|^n} = \omega_{n-1} \log \frac{R_*}{r_*} \tag{1.40}$$

More generally, if $h : \mathbb{X} \xrightarrow{\text{onto}} \mathbb{A}^*$ is conformal then the modulus of $\mathbb{X}$ is defined by pulling back the measure $\mathrm{d}\mu$ to $\mathbb{X}$.

$$\operatorname{Mod}\mathbb{X} = \int_{\mathbb{X}} \frac{J(x,h)\,\mathrm{d}x}{|h(x)|^n} = n^{-\frac{n}{2}} \int_{\mathbb{X}} \frac{\|\,Dh(x)\,\|^n\,\mathrm{d}x}{|h(x)|^n} \tag{1.41}$$

Obviously, this definition is free from the choice of the conformal map h. For $n = 2$ the modulus of a doubly connected domain is the only conformal invariant; that is, the Riemann moduli space is one dimensional. The situation is much more rigid for $n \geqslant 3$. If now $h : \mathbb{X} \xrightarrow{\text{onto}} \mathbb{A}^*$ is any permissible deformation, then

$$\mathscr{F}_h \stackrel{\text{def}}{=\!=} \int_{\mathbb{X}} \frac{\|\,Dh(x)\,\|^n\,\mathrm{d}x}{|h(x)|^n} \geqslant n^{\frac{n}{2}} \int_{\mathbb{X}} \frac{J(x,h)\,\mathrm{d}x}{|h(x)|^n} = n^{\frac{n}{2}} \operatorname{Mod}\mathbb{A}^* \tag{1.42}$$

Naturally, this integral tells us about how much h differs from a conformal mapping in an average sense. On the other hand, Quasiconformal Theory deals with pointwise distortions. Among them is the *outer distortion function*

$$\mathcal{K}_O(x,h) \stackrel{\text{def}}{=\!=} \begin{cases} \frac{\|Dh(x)\|^n}{n^{\frac{n}{2}} J(x,h)}, & \text{if } J(x,h) > 0 \\ 1, & \text{otherwise} \end{cases} \tag{1.43}$$

Let us push forward $\mathcal{K}_O$ to the target space via the mapping h itself, so as to obtain a function

$$\mathcal{K}_h(y) \stackrel{\text{def}}{=\!=} \mathcal{K}_O(x,h), \qquad \text{where } x = h^{-1}(y) \tag{1.44}$$

We note, without proof, that

$$\mathscr{F}_h = n^{\frac{n}{2}} \int_{\mathbb{Y}} \mathcal{K}_h(y)\, \mathrm{d}\mu(y) \tag{1.45}$$

This discussion leads us to the minimization of the $\mathscr{L}^1(\mathbb{Y}, \mathrm{d}\mu)$-norm of the outer distortion. One might suspect that the minimum will be attained when the distortion function is constant. Indeed, when the domain is also an annulus, say $\mathbb{X} = \mathbb{A} = \mathbb{A}(r,R)$, then the power stretching

$$h^\alpha(x) = \lambda |x|^{\alpha-1} x, \qquad \text{where } \alpha = \frac{\operatorname{Mod}\mathbb{A}^*}{\operatorname{Mod}\mathbb{A}} \text{ and } \lambda = r_* r^{-\alpha} = R_* R^{-\alpha} \tag{1.46}$$

has constant outer distortion $\mathcal{K}_O(x,h^\alpha) = \alpha^{-1} n^{-\frac{n}{2}} (\alpha^2 + n - 1)^{\frac{n}{2}}$. Nevertheless, it takes some effort to show that in dimensions $n = 2, 3$, h^α is actually a minimizer of $\mathscr{F}_h$ among all homeomorphisms.

THEOREM 1.14. *Let $\mathbb{A}$ and $\mathbb{A}^*$ be spherical annuli in $\mathbb{R}^n$, $n = 2,3$. Then for every homeomorphism $h\colon \mathbb{A} \xrightarrow{\text{onto}} \mathbb{A}^*$ we have*

$$\mathscr{F}_h \stackrel{\text{def}}{=\!=} \int_{\mathbb{A}} \frac{\|Dh\|^n}{|h|^n} \geqslant \left(n - 1 + \alpha^2\right)^{\frac{n}{2}} \operatorname{Mod}\mathbb{A}, \qquad \text{where } \alpha = \frac{\operatorname{Mod}\mathbb{A}^*}{\operatorname{Mod}\mathbb{A}} \tag{1.47}$$

Equality holds for the power stretching $h(x) = r_* r^{-\alpha}|x|^{\alpha-1}x$, uniquely up to conformal automorphisms of $\mathbb{A}$.

It is somewhat surprising that in dimensions $n \geqslant 4$, this feature of h^α no longer holds when the target annulus is conformally too fat. We have the following result.

THEOREM 1.15. *For each $n \geqslant 4$, there exists $\alpha_n > 1$ such that (1.47) holds whenever*

$$\alpha \stackrel{\text{def}}{=\!=} \frac{\operatorname{Mod}\mathbb{A}^*}{\operatorname{Mod}\mathbb{A}} < \alpha_n \tag{1.48}$$

The power stretching $h(x) = r_* r^{-\alpha}|x|^{\alpha-1}x$ is the only minimizer of $\mathscr{F}_h$ modulo conformal automorphism of $\mathbb{A}$.

Examples will be given to show that the extremals are no longer power stretchings if

$$\frac{\operatorname{Mod}\mathbb{A}^*}{\operatorname{Mod}\mathbb{A}} \geqslant \sqrt{\frac{n-1}{n-3}} \tag{1.49}$$

In other words, the upper bound for α in (1.48) lies in the interval $1 < \alpha_n < \sqrt{\frac{n-1}{n-3}}$. See Chapter 12 for more precise estimates of α_n. Moreover, if $\operatorname{Mod}\mathbb{A}^*$ is too large relative to $\operatorname{Mod}\mathbb{A}$, then the extremals cannot be found even within general radial mappings, see Chapter 14.

14. Free Lagrangians

In 1977 a novel approach towards minimization of polyconvex energy functionals was developed and published by J. Ball [**5**]. The underlying idea was to view the integrand as convex function of null Lagrangians. The term null Lagrangian pertains to a nonlinear differential expression whose integral over any open region depends only on the boundary values of the map, like integrals of an exact differential form. The interested reader is referred to [**6, 15, 22**]. But we are concerned with mappings $h : \mathbb{X} \to \mathbb{Y}$ that are free on the boundary. The only condition we impose on h is that it is a weak $\mathscr{W}^{1,n}$-limit of homeomorphisms from $\mathbb{X}$ onto $\mathbb{Y}$. There still exist some nonlinear differential forms, associated with a given pair of domains $\mathbb{X}$ and $\mathbb{Y}$, whose integral means over $\mathbb{X}$ remain the same within a given class of deformations $h\colon \mathbb{X} \xrightarrow{\text{onto}} \mathbb{Y}$, regardless of their boundary values. These are rather special null Lagrangians. The simplest example is furnished by the Jacobian determinant of an orientation preserving homeomorphism $h \in \mathscr{W}^{1,n}(\mathbb{X}, \mathbb{Y})$

$$\int_{\mathbb{X}} J(x,h)\,\mathrm{d}x = |\mathbb{Y}|. \tag{1.50}$$

One might ask which expressions of the type $E(x, h, Dh)$ enjoy identities such as this? We call them *free Lagrangians*. Such a notion lies fairly deep in the topology of the mappings $h : \mathbb{X} \to \mathbb{Y}$. For example, using the topological degree, we find that the differential expression

$$E(x,h,Dh)\,\mathrm{d}x = \sum_{i=1}^{n} \frac{h^i\, dh^1 \wedge \ldots \wedge dh^{i-1} \wedge d|x| \wedge dh^{i+1} \wedge \ldots \wedge dh^n}{|x|\,|h|^n} \tag{1.51}$$

is a free Lagrangian within the class of orientation preserving homeomorphisms $h : \mathbb{A} \to \mathbb{A}^*$ between annuli. Indeed, we have the desired identity

$$\int_{\mathbb{A}} E(x,h,Dh)\,\mathrm{d}x \;=\; \operatorname{Mod}\mathbb{A} \qquad \text{for every } \; h \in \mathscr{P}(\mathbb{X},\mathbb{Y}) \tag{1.52}$$

A peculiarity of this example is further emphasized by the fact that the target annulus $\mathbb{A}^*$ does not even enter into this identity.

Like in the theory of polyconvex energies, the minimization of an energy integral whose integrand is a convex function of a number of free Lagrangians poses no challenge; Jensen's inequality usually gives the desired sharp lower bounds. Unfortunately, our integrand $\| Dh \|^n$ cannot be expressed as a convex function of free Lagrangians; though it is a convex function of the usual null Lagrangians.

15. Uniqueness

To reach the uniqueness conclusions, we first show that any extremal mapping $h : \mathbb{A} \to \mathbb{A}^*$ satisfies the following system of nonlinear PDEs

$$D^*h(x)\, Dh(x) = \mathbf{G}(x,h) \tag{1.53}$$

where the Cauchy-Green tensor $\mathbf{G}$ actually depends only on two variables $|x|$ and $|h|$. As this system is overdetermined it comes as no surprise that any two solutions $h^\circ(x)$ and $h(x)$ for which $|h^\circ(a)| = |h(a)|$, at some $a \in \mathbb{A}$, must be equal modulo an orthogonal transformation $T : \mathbb{R}^n \to \mathbb{R}^n$, namely $h(x) = Th^\circ(x)$. Our proof exploits the classical computation of curvature of $\mathbf{G}$ in terms of its Christoffel symbols. As regards the existence of such point $a \in \mathbb{A}$, we shall again rely on estimates of free Lagrangians.

16. The $\mathscr{L}^1$-theory of inner distortion

The conformal energy can naturally be turned around so as to yield, the $\mathscr{L}^1$-integrability of a distortion function of the inverse mapping [**4**].

We recall that a homeomorphism $f : \mathbb{Y} \to \mathbb{X}$ of Sobolev class $\mathscr{W}^{1,1}_{\mathrm{loc}}(\mathbb{Y},\mathbb{X})$ is said to have finite distortion if there exists a measurable function $1 \leqslant K(y) < \infty$ such that

$$|Df(y)|^n \leqslant K(y)\, J(y,f) \tag{1.54}$$

for almost every $y \in \mathbb{Y}$. Here $|Df|$ stands for the operator norm of Df. Using the Hilbert-Schmidt norm of matrices the outer distortion function of f takes the form

$$\mathbb{K}_O(y,f) = \frac{\| Df(y) \|^n}{n^{\frac{n}{2}} J(y,f)} \tag{1.55}$$

if $J(y,f) > 0$ and we set $\mathbb{K}_O(y,f) = 1$, otherwise. There are many more distortion functions of great importance in Geometric Function Theory. Among them are the inner distortion functions.

$$K_I(y,f) = \frac{|D^\sharp f(y)|^n}{\det D^\sharp f(y)}\ ^{4} \tag{1.56}$$

and

$$\mathbb{K}_I(y,f) = \frac{\| D^\sharp f(y) \|^n}{n^{\frac{n}{2}} \det D^\sharp f(y)} \tag{1.57}$$

at the points where $J(y,f) > 0$. Otherwise, we set

$$\mathbb{K}_I(y,f) = K_I(y,f) = 1 \tag{1.58}$$

In recent years there has been substantial interest in mappings with integrable distortion [**2, 4, 20, 23, 32**]. Suppose now that $f \in \mathscr{W}^{1,n-1}_{\mathrm{loc}}(\mathbb{Y},\mathbb{R}^n)$. The following identities connect the n-harmonic integrals with the theory of mappings with integrable inner distortion

$$\int_{\mathbb{Y}} K_I(y,f)\,\mathrm{d}y = \int_{\mathbb{X}} |Dh(x)|^n\,\mathrm{d}x \tag{1.59}$$

and

$$n^{\frac{n}{2}} \int_{\mathbb{Y}} \mathbb{K}_I(y,f)\,\mathrm{d}y = \int_{\mathbb{X}} \| Dh(x) \|^n\,\mathrm{d}x \tag{1.60}$$

where h denotes the inverse of f, [**12, 19, 20**]. These identities imply that the inverse of a mapping of integrable distortion always lies in the Sobolev space $h \in \mathscr{W}^{1,n}(\mathbb{X},\mathbb{Y})$. Similarly, the $\mathscr{L}^1(\mathbb{A}^*,\mathrm{d}\mu)$-integral means of $K_I(y,f)$ and $\mathbb{K}_I(y,f)$ with respect to the dimensionless weight $\mathrm{d}\mu = |y|^{-n}\,\mathrm{d}y$ are the energy functionals $\mathscr{F}_h$:

$$\int_{\mathbb{A}^*} \frac{K_I(y,f)}{|y|^n}\,\mathrm{d}y = \int_{\mathbb{A}} \frac{|Dh(x)|^n}{|h(x)|^n}\,dx \tag{1.61}$$

and

$$n^{\frac{n}{2}} \int_{\mathbb{A}^*} \frac{\mathbb{K}_I(y,f)}{|y|^n}\,\mathrm{d}y = \int_{\mathbb{A}} \frac{\| Dh(x) \|^n}{|h(x)|^n}\,dx \tag{1.62}$$

Returning to (1.60), we infer from minimal n-harmonic mappings that

[4]Here we use the operator norm of the cofactor matrix $D^\sharp f$.

THEOREM 1.16. *Under the Nitsche bounds*

$$\aleph_{\dagger}(\operatorname{Mod}\mathbb{A}) \leqslant \operatorname{Mod}\mathbb{A}^* \leqslant \aleph^{\dagger}(\operatorname{Mod}\mathbb{A}) \tag{1.63}$$

the $\mathscr{L}^1(\mathbb{Y})$-norm of the inner distortion $\mathbb{K}_I(y,f)$ assumes its minimum value on a mapping $f\colon \mathbb{A}^* \xrightarrow{\text{onto}} \mathbb{A}$ whose is inverse is a radial n-harmonic mapping $h^\circ\colon \mathbb{A} \xrightarrow{\text{onto}} \mathbb{A}^*$. Such an extremal mapping f is unique up to a conformal automorphism of $\mathbb{A}$.

THEOREM 1.17. *If the domain annulus $\mathbb{A}^*$ is too thin relative to the target annulus $\mathbb{A}$; precisely, under the condition*

$$\operatorname{Mod}\mathbb{A}^* \;<\; \aleph_{\dagger}(\operatorname{Mod}\mathbb{A}) \qquad \textit{-below the Nitsche bound.} \tag{1.64}$$

then the infimum of the $\mathscr{L}^1(\mathbb{A}^*)$-norm of $\mathbb{K}_I(y,f)$ is not attained among homeomorphisms $f\colon \mathbb{A}^* \xrightarrow{\text{onto}} \mathbb{A}$.

Nevertheless, we were able to find the infimum of the $\mathscr{L}^1$-norms and identify the minimizing sequences. The weak BV-limits of such sequences and the underlying concept of their distortion (to be defined) are worth carrying out.

Conclusion

At the first glance the problems we study here may appear to be merely technical. However, their solutions require truly innovative approaches with surprising outcomes. For instance, the case $n \geqslant 4$ is different than one might a priori expect; upper bounds for the modulus of $\mathbb{A}^*$ are necessary in order to ensure radial symmetry of the minimizers. The underlying technique of integration of various nonlinear differential forms is interesting in its own right. The *free Lagrangians* play a pivotal role, like null Lagrangians did play in the polyconvex calculus of variations. The entire subject grew out of fundamental questions of Quasiconformal Geometry about mappings of integrable distortion. The paper also embraces a number of important aspects of the Calculus of Variation.

We present this little theory here in two parts.

Part 1

Principal Radial n-Harmonics

CHAPTER 2

Nonexistence of n-Harmonic Homeomorphisms

That nice smooth domains $\mathbb{X}$ and $\mathbb{Y}$, such as annuli, may not admit a homeomorphism $h\colon \mathbb{X} \xrightarrow{\text{onto}} \mathbb{Y}$ of smallest conformal energy is a sequel of even more general observation. Let $\mathcal{A}$ be a topological annulus in $\mathbb{R}^n$, we consider all possible n-harmonic homeomorphisms $h\colon \mathcal{A} \to \mathbb{R}^n$. Unconcerned about the energy (finite or infinite) we address the question; which topological annuli $\mathcal{A}^*$ can be obtained as images of $\mathcal{A}$ under such mappings? In general, this problem lies beyond our methods. Even in dimension $n = 2$ and when $\mathcal{A}^*$ is a circular annulus (the Nitsche conjecture) the answer to this question required rather sophisticated ideas [**26**]. Nevertheless, one may roughly say that $\mathcal{A}^*$ cannot be very thin. A specific instance is as follows. Let $\mathbb{S}$ be a convex $(n-1)$-hypersurface in $\mathbb{R}^n$, given by

$$\mathbb{S} = \{y \in \mathbb{R}^n \colon F(y) = 0\},$$

where F is a $\mathscr{C}^2$-smooth function in a neighborhood of $\mathbb{S}$, such that $\nabla F \neq 0$ on $\mathbb{S}$. We assume that the Hessian matrix

$$\nabla^2 F = \left[\frac{\partial^2 F}{\partial y_i\, \partial y_j}\right] \quad \text{is positive definite on } \mathbb{S}.$$

Consider the ϵ-vicinity of $\mathbb{S}$; that is, $\mathbb{V}_\epsilon = \{y \colon |F(y)| < \epsilon\}$.

PROPOSITION 2.1. *If ϵ is small enough then there is no n-harmonic homeomorphism $h\colon \mathcal{A} \xrightarrow{\text{into}} \mathbb{V}_\epsilon$ such that $\mathbb{S} \subset h(\mathcal{A})$.*

PROOF. We shall not give any explicit bound for ϵ, but instead we argue by contradiction. Suppose that for every positive integer ℓ, thus $\epsilon = 1/\ell$, there exists an n-harmonic homeomorphism

$$h_\ell = (h_\ell^1, \dots, h_\ell^n) \colon \mathcal{A} \xrightarrow{\text{onto}} h_\ell(\mathcal{A}), \quad \text{such that} \quad \mathbb{S} \subset h_\ell(\mathcal{A}) \subset \mathbb{V}_\epsilon, \quad \epsilon = \frac{1}{\ell}.$$

We appeal to the $\mathscr{C}^{1,\alpha}$-regularity theory of n-harmonic mappings [**13, 36, 47, 48, 49**]. Accordingly, h_ℓ are uniformly bounded in $\mathscr{C}^{1,\alpha}$-norm on every compact subset of $\mathcal{A}$. Moreover, there is a subsequence, again denoted by $\{h_\ell\}$, which converges together with the first order derivatives to a mapping $h\colon \mathcal{A} \to \mathbb{S}$ uniformly on compact subsets. The limit mapping $h = (h^1, \dots, h^n)$ still satisfies the n-harmonic equation. We aim to show that $Dh \equiv 0$ in $\mathcal{A}$, meaning that h is constant. Recall that $Dh\colon \mathcal{A} \to \mathbb{R}^{n\times n}$ is (Hölder) continuous. The computation below is certainly valid in the open region where $Dh \neq 0$, because h is $\mathscr{C}^\infty$-smooth in such region. We have

$$F\big(h(x)\big) = 0 \qquad \text{for all } x \in \mathcal{A}.$$

Applying partial differentiation $\frac{\partial}{\partial x_\nu}$ and chain rule yields

$$\sum_{i=1}^{n} \frac{\partial F}{\partial y_i}\frac{\partial h^i}{\partial x_\nu} = 0, \qquad \text{for } \nu = 1, 2, \ldots, n. \tag{2.1}$$

On the other hand, the n-harmonic equation takes the form

$$\sum_{\nu=1}^{n} \frac{\partial}{\partial x_\nu}\left[\lambda^{n-2}\frac{\partial h^i}{\partial x_\nu}\right] = 0 \quad \text{for } i = 1, 2, \ldots, n, \quad \text{where } \lambda = \| Dh(x) \| \tag{2.2}$$

or, equivalently

$$\sum_{\nu=1}^{n}\left[(n-2)\lambda^{n-3}\frac{\partial \lambda}{\partial x_\nu}\frac{\partial h^i}{\partial x_\nu} + \lambda^{n-2}\frac{\partial^2 h^i}{\partial x_\nu \partial x_\nu}\right] = 0 \quad i = 1, \ldots, n. \tag{2.3}$$

We multiply these equations by $\frac{\partial F}{\partial y_i}$, sum them up with respect to i, and use the identity (2.1) to obtain

$$\lambda^{n-2}\sum_{i=1}^{n}\sum_{\nu=1}^{n}\frac{\partial F}{\partial y_i}\frac{\partial^2 h^i}{\partial x_\nu \partial x_\nu} = 0. \tag{2.4}$$

Next we differentiate (2.1) with respect to ν and sum up the equations,

$$\sum_{i=1}^{n}\sum_{\nu=1}^{n}\left(\frac{\partial F}{\partial y_i}\frac{\partial^2 h^i}{\partial x_\nu \partial x_\nu} + \sum_{j=1}^{n}\frac{\partial^2 F}{\partial y_i \partial y_j}\frac{\partial h^j}{\partial x_\nu}\frac{\partial h^i}{\partial x_\nu}\right) = 0. \tag{2.5}$$

Here the double sum for the first term in (2.5) vanishes due to (2.4), so we have

$$\sum_{\nu=1}^{n}\left(\sum_{i,j}\frac{\partial^2 F}{\partial y_i \partial y_j}\frac{\partial h^i}{\partial x_\nu}\frac{\partial h^j}{\partial x_\nu}\right) = 0.$$

Since the Hessian matrix of F is positive definite, this equation yields $\frac{\partial h^i}{\partial x_\nu} = 0$ for all i, $\nu = 1, 2, \ldots, n$, as desired.

To reach a contradiction we look more closely at the homeomorphisms $h_\ell \colon \mathcal{A} \overset{\text{onto}}{\longrightarrow} h_\ell(\mathcal{A})$. Choose and fix an $(n-1)$-dimensional hypersurface $\Sigma \subset \mathcal{A}$ (topological sphere) which separates the boundary components of $\mathcal{A}$. Its images $h_\ell(\Sigma) \subset h_\ell(\mathcal{A})$ separate the boundary components of $h_\ell(\mathcal{A})$. Thus, in particular,

$$0 < \liminf_{\ell \to \infty} \, [\operatorname{diam} h_\ell(\Sigma)] = \operatorname{diam} h(\Sigma) = 0$$

a clear contradiction. □

It is not so clear, however, how thick $\mathcal{A}^*$ should be to ensure existence of n-harmonic homeomorphisms $h \colon \mathcal{A} \overset{\text{onto}}{\longrightarrow} \mathcal{A}^*$. In dimension $n = 2$ the condition $\operatorname{Mod} \mathcal{A}^* \geqslant \operatorname{Mod} \mathcal{A}$ is sufficient [**24**] and is not far from being necessary [**27**]. Questions of existence of harmonic diffeomorphisms between surfaces are treated in [**33**].

CHAPTER 3

Generalized n-Harmonic Mappings

In the classical Dirichlet problem one asks for the energy minimal mapping $h\colon \mathbb{X} \to \mathbb{R}^n$ of the Sobolev class $h \in h_\circ + \mathscr{W}_\circ^{1,n}(\mathbb{X}, \mathbb{R}^n)$ whose boundary values are explicitly prescribed by means of a given mapping $h_\circ \in \mathscr{W}^{1,n}(\mathbb{X}, \mathbb{R}^n)$. We refer to the works of C. B. Morrey [**38, 39**], where a systematic study of variational methods in the theory of harmonic integrals in vectorial case originated, see also [**40**]. The variation $h \rightsquigarrow h + \epsilon\eta$, in which $\eta \in \mathscr{C}_\circ^\infty(\mathbb{X}, \mathbb{R}^n)$ and $\epsilon \to 0$, leads to the integral form of the familiar n-harmonic system of equations

$$\int_{\mathbb{X}} \langle \| Dh \|^{n-2} Dh,\, D\eta \rangle = 0, \qquad \text{for every } \eta \in \mathscr{C}_\circ^\infty(\mathbb{X}, \mathbb{R}^n). \tag{3.1}$$

Equivalently

$$\Delta_n h = \operatorname{Div}\big(\| Dh \|^{n-2} Dh \big) = 0, \quad \text{in the sense of distributions} \tag{3.2}$$

or, entry-wise, for $h = (h^1, \dots, h^n)$

$$\sum_{i=1}^{n} \big(\| Dh \|^{n-2} h^\alpha_{x_i} \big)_{x_i} = 0, \quad \alpha = 1, 2, \dots, n. \tag{3.3}$$

In nonlinear elastostatics the matrix field

$$\mathcal{S} = \mathcal{S}h = \| Dh(x) \|^{n-2} Dh(x)$$

is known as the *Piola-Kirchoff* tensor for the energy density function $W(x) = \| Dh(x) \|^n$. This tensor represents the stress induced by $h\colon \mathbb{X} \to \mathbb{R}^n$. Mappings of Sobolev class $\mathscr{W}^{1,n}_{\text{loc}}(\mathbb{X}, \mathbb{R}^n)$ that satisfy the equation $\operatorname{Div}\mathcal{S}h = 0$ are called the *equilibrium solutions.* Equilibrium solutions in a given class $h_\circ + \mathscr{W}_\circ^{1,n}(\mathbb{X}, \mathbb{R}^n)$ represent unique minimizers within this class. The situation is dramatically different if we allow h to slip freely along the boundaries. The *inner variation* works well in this case. This is simply a change of the independent variable; $h_\epsilon = h \circ \eta_\epsilon$, where $\eta_\epsilon \colon \mathbb{X} \xrightarrow{\text{onto}} \mathbb{X}$ are $\mathscr{C}^\infty$-smooth automorphisms of $\mathbb{X}$ onto itself, depending smoothly on a parameter $\epsilon \approx 0$ where $\eta_\circ = id\colon \mathbb{X} \xrightarrow{\text{onto}} \mathbb{X}$. Let us take on, as a initial step, the inner variation of the form

$$\eta_\epsilon(x) = x + \epsilon\, \eta(x), \qquad \eta \in \mathscr{C}_\circ^\infty(\mathbb{X}, \mathbb{R}^n). \tag{3.4}$$

We compute

$$\begin{aligned} Dh_\epsilon(x) &= Dh(x + \epsilon\eta)(I + \epsilon D\eta) \\ \| Dh_\epsilon(x) \|^n &= \| Dh \|^n + n\epsilon \, \langle \| Dh \|^{n-2} D^*h \cdot Dh \,,\, D\eta \rangle + o(\epsilon). \end{aligned}$$

Be aware that in this equation Dh is evaluated at the point $y = x + \epsilon\, \eta(x) \in \mathbb{X}$. Integration with respect to x-variable yields a formula for the energy of h_ϵ,

$$\mathscr{E}_{h_\epsilon} = \int_{\mathbb{X}} \big[\| Dh \|^n + n\epsilon \langle \| Dh \|^{n-2} D^*h \cdot Dh \,,\, D\eta \rangle \big] \, dx + o(\epsilon).$$

We now make the substitution $y = x + \epsilon\,\eta(x)$ for which the following transportation rules apply: $x = y - \epsilon\,\eta(y) + o(\epsilon)$, $\eta(x) = \eta(y) + o(1)$ and the change of volume element $\mathrm{d}x = [1 - \epsilon \operatorname{Tr} D\eta(y)]\,\mathrm{d}y + o(\epsilon)$. The equilibrium equation for the inner variation is obtained from $\frac{\mathrm{d}}{\mathrm{d}\epsilon}\mathscr{E}_{h_\epsilon} = 0$ at $\epsilon = 0$,

$$\int_{\mathbb{X}} \langle \| Dh \|^{n-2} D^*h \cdot Dh - \frac{1}{n} \| Dh \|^{n} I \,,\, D\eta \rangle \,\mathrm{d}y = 0 \tag{3.5}$$

or, by means of distributions

$$\operatorname{Div}\left(\| Dh \|^{n-2} D^*h \cdot Dh - \frac{1}{n} \| Dh \|^{n} I \right) = 0. \tag{3.6}$$

Now we introduce the divergence free tensor

$$\Lambda = \| Dh \|^{n-2} D^*h \cdot Dh - \frac{1}{n} \| Dh \|^{n} I = \left(C - \frac{1}{n}\operatorname{Tr} C \right) \operatorname{Tr}^{\frac{n-2}{n}} C$$

where we recall the right Cauchy-Green tensor $C = C(x) = D^*h \cdot Dh$. The name *generalized n-harmonic equation* will be given to (3.6) because of the following:

LEMMA 3.1. *Every n-harmonic mapping $h \in \mathscr{W}^{1,n}_{loc}(\mathbb{X}, \mathbb{R}^n)$ solves the generalized n-harmonic equation* (3.6).

PROOF. We consider again the perturbed mappings $h_\epsilon(x) = h(x + \epsilon\,\eta(x))$ which coincide with h outside a subdomain $\mathbb{U}$ compactly contained in $\mathbb{X}$. Applying the integral form (3.1) of the n-harmonic equation, but with the test function $h - h_\epsilon \in \mathscr{W}^{1,n}_\circ(\mathbb{U}, \mathbb{R}^n)$ in place of η, [1] we estimate the energy of h over $\mathbb{U}$ as follows

$$\begin{aligned}
\mathscr{E}_h &= \int_{\mathbb{U}} \| Dh \|^{n} = \int_{\mathbb{U}} \| Dh \|^{n-2} \langle Dh \,,\, Dh \rangle \\
&= \int_{\mathbb{U}} \| Dh \|^{n-2} \langle Dh \,,\, Dh_\epsilon \rangle \leqslant \left(\int_{\mathbb{U}} \| Dh \|^{n} \right)^{\frac{n-1}{n}} \left(\int_{\mathbb{U}} \| Dh_\epsilon \|^{n} \right)^{\frac{1}{n}} .
\end{aligned}$$

Hence

$$\mathscr{E}_h \leqslant \mathscr{E}_{h_\epsilon}, \qquad \text{with equality at } \epsilon = 0.$$

This means $\frac{\mathrm{d}}{\mathrm{d}\epsilon}\mathcal{E}_{h_\epsilon} = 0$ at $\epsilon = 0$. Equivalently, h satisfies the generalized n-harmonic equation. □

In dimension $n = 2$, the generalized harmonic equation reduces to

$$\operatorname{Div}\left(D^*h\, Dh - \frac{1}{2} \| Dh \|^{2} I \right) = 0. \tag{3.7}$$

This equation is known as Hopf-Laplace equation [**11**]. In complex notation it takes the form

$$\frac{\partial}{\partial \bar z}\left(h_z \overline{h_{\bar z}} \right) = 0, \qquad z = x_1 + ix_2. \tag{3.8}$$

If, by some reason, $h \in \mathscr{C}^2(\mathbb{X}, \mathbb{C})$ then (3.8) reads as

$$J(z,h)\Delta h = 0, \qquad \text{where } \Delta = 4\frac{\partial^2}{\partial z \partial \bar z}. \tag{3.9}$$

There exist diverse non-harmonic solutions to (3.9), many of them of great interest in the theory of minimal surfaces and some with potential applications to nonlinear elasticity (elastic plates), see [**11**].

[1] This is justified because $\mathscr{C}^\infty_\circ(\mathbb{U}, \mathbb{R}^n)$ is dense in $\mathscr{W}^{1,n}_\circ(\mathbb{U}, \mathbb{R}^n)$.

Equally in higher dimensions, the n-harmonic mappings of the Sobolev class $\mathscr{W}^{1,n}_{\mathrm{loc}}(\mathbb{X},\mathbb{R}^n)$ are only particular solutions to the generalized n-Laplacian.

1. Solutions to the generalized n-harmonic equation that are not n-harmonic

Let us take on stage a radial mapping

$$h(x) = H(|x|)\frac{x}{|x|}, \qquad \text{where } H = H(t) \text{ is absolutely continuous}$$

see Chapter 4 for extensive treatment of such mappings. We find that

$$\begin{aligned}\Lambda &= \| Dh \|^{n-2}\left(D^*h\cdot Dh - \frac{1}{n}\| Dh \|^2 I\right)\\ &= (n-1)^{\frac{n-2}{n}}\left(H^2 + \frac{|x|^2\dot{H}^2}{n-1}\right)^{\frac{n-2}{2}}\left(H^2 - |x|^2\dot{H}^2\right)\frac{1}{|x|^n}\left(\frac{x\otimes x}{|x|^2} - \frac{1}{n}I\right).\end{aligned} \tag{3.10}$$

It is shown in Chapter 5, that if h is a $\mathscr{C}^2$-smooth n-harmonic mapping then $H = H(t)$ must satisfy the *characteristic equation*

$$\left(H^2 + \frac{|x|^2\dot{H}^2}{n-1}\right)^{\frac{n-2}{2}}\cdot\left(H^2 - |x|^2\dot{H}^2\right) \equiv \text{const.} \tag{3.11}$$

This also confirms that h satisfies the generalized n-harmonic equation because of the identity

$$\operatorname{Div}\frac{1}{|x|^n}\left(\frac{x\otimes x}{|x|^2} - \frac{1}{n}I\right) = 0.$$

However, the *hammering mapping* $h(x) = \frac{x}{|x|}$, corresponding to $H(t)\equiv 1$, solves equation (3.6) but is not n-harmonic. If we paste this hammering mapping with a compatible smooth radial n-harmonic solution (the critical Nitsche mapping) there will emerge a $\mathscr{C}^{1,1}$-smooth minimal deformation between annuli. There are many more non n-harmonic solutions to (3.6), some seem to be unsatisfactory; for example those with Jacobian changing sign are forbidden in elasticity theory because of the principle of non-interpenetration of matter.

2. Slipping along the boundaries

Let us return to the inner variation in (3.4), $h_\epsilon(x) = h(x+\epsilon\,\eta(x))$, but this time with η not necessarily having compact support. We assume that $\mathbb{X}$ is a $\mathscr{C}^1$-smooth domain and $\eta\in\mathscr{C}^1(\overline{\mathbb{X}})$. The fact that h is allowed to freely slip along the boundaries amounts to saying that the vector field $\eta=\eta(x)$ is tangent to $\partial\mathbb{X}$ at every point $x\in\partial\mathbb{X}$. The integral form of the resulting variational equation is the same as in (3.5). Integration by parts (Green's formula) will produce no integral over $\mathbb{X}$, because of (3.6); there will remain only boundary integrals. Precisely, a general formula we are referring to is:

$$0 = \int_{\mathbb{X}}\langle\Lambda, D\eta\rangle\,\mathrm{d}x = -\int_{\mathbb{X}}\langle\operatorname{Div}\Lambda,\eta\rangle\,\mathrm{d}x + \int_{\partial\mathbb{X}}\langle\Lambda(x)\mathfrak{n}(x),\eta(x)\rangle\,\mathrm{d}\sigma(x)$$

where $\mathfrak{n}(x)$ is the outer unit vector field to $\partial\mathbb{X}$ and $\mathrm{d}\sigma(x)$ stands for the surface measure. This is justified under an appropriate assumption on the degree of integrability of Λ, DivΛ, η and $D\eta$. Nevertheless in our case, we obtain

$$\int_{\partial\mathbb{X}} \langle (\,\|Dh\|^{n-2} D^*h\cdot Dh - \frac{1}{n}\|Dh\|^n I)\mathfrak{n}(x)\,,\,\eta(x)\rangle\,\mathrm{d}\sigma(x) = 0 \tag{3.12}$$

Since $\eta(x)$ can be any tangent field the equation (3.12) is possible if and only if the vector field

$$\left(\|Dh\|^{n-2} D^*h\cdot Dh - \frac{1}{n}\|Dh\|^n I\right)\mathfrak{n}(x)$$

is orthogonal to $\partial\mathbb{X}$. In other words, at each $x \in \partial\mathbb{X}$ the linear mapping $\|Dh\|^{n-2}D^*h\cdot Dh - \frac{1}{n}\|Dh\|^n I$ takes the (one-dimensional) space of normal vectors into itself. Of course, the same holds for the mapping $D^*h\cdot Dh$. Since $D^*h\cdot Dh$ is symmetric one can say, equivalently, that $D^*h\cdot Dh$ preserves the tangent space. We just proved the following.

PROPOSITION 3.2. *The equilibrium solution for mappings that are slipping along the boundaries satisfies, in addition to* (3.6)*, the following condition*

$$D^*h\cdot Dh\colon \mathbf{T}_x\partial\mathbb{X} \to \mathbf{T}_x\partial\mathbb{X} \tag{3.13}$$

equivalently,

$$D^*h\cdot Dh\colon \mathbf{N}_x\partial\mathbb{X} \to \mathbf{N}_x\partial\mathbb{X}. \tag{3.14}$$

where $\mathbf{T}_x\partial\mathbb{X}$ *and* $\mathbf{N}_x\partial\mathbb{X}$ *designate the tangent and normal spaces at* $x \in \partial\mathbb{X}$.

In dimension $n = 2$ this amounts to saying that the Hopf quadratic differential $h_z\overline{h_{\bar z}}\,\mathrm{d}z^2$ is real along the boundary components of $\mathbb{X}$. In an annulus $\mathbb{A} = \{z\colon r < |z| < R\}$ every Hopf differential which is real on $\partial\mathbb{A}$ takes the form

$$h_z\overline{h_{\bar z}} \equiv \frac{c}{z^2}, \qquad c \text{ is a real constant}$$

see [**11**]. One of the solutions is the mapping we are already encountered in (1.20)

$$h(z) = \begin{cases} \frac{z}{|z|} & \frac{1}{R} < |z| \leqslant 1\,, \qquad \text{hammering part} \\ \frac{1}{2}\left(z + \frac{1}{\bar z}\right) & 1 \leqslant |z| \leqslant R\,, \qquad \text{harmonic part} \end{cases}$$

In higher dimensions the radial mapping $h(x) = H\big(|x|\big)\frac{x}{|x|}$ in the annulus $\mathbb{A} = \{x\colon r < |x| < R\}$ also complies with the boundary condition (3.13). Indeed, the Cauchy-Green tensor takes the form

$$D^*h\cdot Dh = \frac{H^2}{|x|^2} I + \left(\dot H^2 - \frac{H^2}{|x|}\right)\frac{x\otimes x}{|x|^2}.$$

The normal vector field at $\partial\mathbb{A}$ is $\mathfrak{n} = \mathfrak{n}(x) = x$, and we see that

$$[D^*h(x)\cdot Dh(x)]\,\mathfrak{n} = \lambda\mathfrak{n}, \qquad \lambda = [\dot H(|x|)]^2.$$

For any tangent vector $\mathfrak{t} = \mathfrak{t}(x)$, we have $(x\otimes x)\mathfrak{t} = 0$ so

$$[D^*h(x)\cdot Dh(x)]\,\mathfrak{t} = \frac{H^2}{|x|^2}\mathfrak{t}.$$

The corresponding singular values of the right Cauchy-Green tensor are $\dot H(x)$ and $\frac{H(|x|)}{|x|}$; *principal stretches.*

Before moving to a study of n-harmonic mappings between spherical annuli, let us fulfill the promise of proving that weak $\mathscr{W}^{1,n}$-limits of homeomorphisms are indeed monotone.

3. Proof of Theorem 1.7

We are dealing with a sequence of homeomorphisms $h_j \colon \mathbb{X} \xrightarrow{\text{onto}} \mathbb{Y}$, $j = 1, 2, \dots$ converging c-uniformly to a mapping $h \colon \mathbb{X} \to \overline{\mathbb{Y}}$ such that

$$\operatorname{dist}\big(h_j(x), \partial\mathbb{Y}\big) \leqslant \eta(x) \, \| \, Dh_j \, \|_{\mathscr{L}^n(\mathbb{X})}.$$

see (1.16). Passing to the limit we also have

$$\operatorname{dist}\big(h(x), \partial\mathbb{Y}\big) \leqslant \eta(x) \liminf_{j\to\infty} \; \| \, Dh_j \, \|_{\mathscr{L}^n(\mathbb{X})}.$$

Therefore, for every $\epsilon > 0$ there is $\delta > 0$ such that

$$\operatorname{dist}\big(h_k(x), \partial\mathbb{Y}\big) < \epsilon \qquad \text{whenever } \operatorname{dist}(x, \partial\mathbb{X}) < \delta \tag{3.15}$$

for every $k = 1, 2, \dots$ and

$$\operatorname{dist}\big(h(x), \partial\mathbb{Y}\big) < \epsilon \qquad \text{whenever } \operatorname{dist}(x, \partial\mathbb{X}) < \delta.$$

This shows that $h^{-1}(\mathbb{K}) \subset \mathbb{X}$ is compact whenever $\mathbb{K} \subset \mathbb{Y}$ is compact. To show that h is also monotone we argue as follows. Consider a continuum $\mathbb{K} \subset \mathbb{Y}$ and assume, to the contrary, that $h^{-1}(\mathbb{K}) = \mathbb{A} \cup \mathbb{B}$, where $\mathbb{A}$ and $\mathbb{B}$ are nonempty disjoint compact subsets of $\mathbb{X}$. For every integer ℓ such that

$$\frac{1}{\ell} < \epsilon \overset{\text{def}}{=\!=} \frac{1}{2}\operatorname{dist}\big(\mathbb{K}, \partial\mathbb{Y}\big)$$

one may consider a neighborhood of $\mathbb{K}$,

$$\mathbb{K}_\ell = \{y \in \mathbb{Y} \colon \operatorname{dist}(y, \mathbb{K}) < 1/\ell\}.$$

This is an open connected set in $\mathbb{Y}$. The preimage $h^{-1}(\mathbb{K}_\ell) \subset \mathbb{X}$ is a neighborhood of the compact set $\mathbb{A} \cup \mathbb{B} \subset \mathbb{X}$. Since $h_j \to h$ c-uniformly then for sufficiently large j, say $j = j_\ell$, we have

$$h_{j_\ell}^{-1}(\mathbb{K}_\ell) \supset \mathbb{A} \cup \mathbb{B}, \qquad \ell \geqslant \frac{1}{\epsilon}.$$

The sets $h_{j_\ell}^{-1}(\mathbb{K}_\ell)$ are connected. Therefore to every $\ell \geqslant \frac{1}{\epsilon}$ there corresponds a point $x_\ell \in h_{j_\ell}^{-1}(\mathbb{K}_\ell) \subset \mathbb{X}$ such that

$$\operatorname{dist}(x_\ell, \mathbb{A}) = \operatorname{dist}(x_\ell, \mathbb{B}). \tag{3.16}$$

We also note that

$$\operatorname{dist}\big(h_{j_\ell}(x_\ell), \partial\mathbb{Y}\big) \geqslant \operatorname{dist}\big(\mathbb{K}_\ell, \partial\mathbb{Y}\big) \geqslant 2\epsilon - \frac{1}{\ell} > \epsilon.$$

Hence, in view of condition (3.15), we have

$$\operatorname{dist}(x_\ell, \partial\mathbb{X}) \geqslant \delta, \qquad \ell = 1, 2, \dots$$

We then choose a subsequence, again denoted by x_ℓ, converging to some point $x \in \mathbb{X}$. Passing to the limit in (3.16) yields

$$\operatorname{dist}(x, \mathbb{A}) = \operatorname{dist}(x, \mathbb{B}), \qquad \text{thus } x \notin \mathbb{A} \cup \mathbb{B}.$$

On the other hand, since h_j converge to h uniformly on compact subsets, we may pass to the limit with the sequence $h_{j_\ell}(x_\ell) \to h(x)$. Finally, since $h_{j_\ell}(x_\ell) \in \mathbb{K}_\ell$ and

$\cap \mathbb{K}_\ell = \mathbb{K}$, we conclude that $h(x) \in \mathbb{K}$, meaning that $x \in h^{-1}(\mathbb{K}) = \mathbb{A} \cup \mathbb{B}$. This contradiction proves Theorem 1.7.

CHAPTER 4

Notation

This chapter is designed to describe basic geometric objects and concepts that will be used throughout this paper. The primary domains here are annuli in $\mathbb{R}^n$, also called spherical rings. They are subsets of the punctured Euclidean space,

$$\mathbb{R}^n_\circ = \mathbb{R}^n \setminus \{0\} = \left\{ x = (x_1, ..., x_n)\,;\ |x| = \sqrt{x_1^2 + \cdots + x_n^2} \neq 0 \right\}, \qquad n \geqslant 2$$

1. Annuli and their modulus

There are four types of annuli, generally denoted by $\mathbb{A} \subset \mathbb{R}^n_\circ$

$$\mathbb{A} = \begin{cases} \mathbb{A}(r,R) = \{x \in \mathbb{R}^n\,;\ r < |x| < R\}\,, & \text{where } 0 \leqslant r < R \leqslant \infty \\ \mathbb{A}[r,R) = \{x \in \mathbb{R}^n\,;\ r \leqslant |x| < R\}\,, & \text{where } 0 < r < R \leqslant \infty \\ \mathbb{A}(r,R] = \{x \in \mathbb{R}^n\,;\ r < |x| \leqslant R\}\,, & \text{where } 0 \leqslant r < R < \infty \\ \mathbb{A}[r,R] = \{x \in \mathbb{R}^n\,;\ r \leqslant |x| \leqslant R\}\,, & \text{where } 0 < r \leqslant R < \infty \end{cases} \tag{4.1}$$

The boundary of each annulus consists of two components (the inner and the outer sphere) except for the degenerate case $\mathbb{A}[r,r]$ which reduces to the $(n-1)$-dimensional sphere denoted by $\mathbb{S}^{n-1}_r$.

The conformal modulus of an annulus is defined by

$$\operatorname{Mod} \mathbb{A} = \omega_{n-1} \log \frac{R}{r} = \int_{\mathbb{A}} \frac{\mathrm{d}x}{|x|^n} \quad \in [0, \infty] \tag{4.2}$$

Hereafter, ω_{n-1} denotes the surface area of the unit sphere $\mathbb{S}^{n-1} \subset \mathbb{R}^n$.

$$\omega_{n-1} = \begin{cases} \frac{2\pi^k}{1\cdot 2\cdots (k-1)}\,, & \text{for } n = 2k \\ \\ \frac{2^{k+1}\pi^k}{1\cdot 3\cdots (2k-1)}\,, & \text{for } n = 2k+1 \end{cases} \tag{4.3}$$

2. Polar coordinates in $\mathbb{R}^n_\circ$

Polar coordinates are natural to use when working with annuli. We associate with any point $x \in \mathbb{R}^n_\circ$ a pair of polar coordinates

$$(r, \omega) \in \mathbb{R}_+ \times \mathbb{S}^{n-1} \sim \mathbb{R}^n_\circ \tag{4.4}$$

where $r = |x|$ is referred to as the radial distance and $\omega = \frac{x}{|x|}$ as the spherical coordinate of x. Obviously $x = r\,\omega$ and the volume element in polar coordinates reads as $\mathrm{d}x = r^{n-1}\,\mathrm{d}r\,\mathrm{d}\omega$

3. Spherical coordinates, latitude and longitude

A position of a point $\omega \in \mathbb{S}^{n-1}$ is determined by its latitude and longitude. These are convenient coordinates for describing deformations of the unit sphere. We refer to the points $\omega_+ = (0, ..., 0, 1)$ and $\omega_- = (0, ..., 0, -1)$ as north and south poles, respectively. The unit sphere with poles removed will be denoted by $\mathbb{S}^{n-1}_{\pm}$. The equatorial sphere $\mathbb{S}^{n-2} \subset \mathbb{S}^{n-1}$ is given by the equations $x_1^2 + ... + x_{n-1}^2 = 1$ and $x_n = 0$. To every point $\omega \in \mathbb{S}^{n-1}_{\pm}$ there correspond meridian coordinates

$$(\theta, \mathfrak{s}) \in (0, \pi) \times \mathbb{S}^{n-2} \sim \mathbb{S}^{n-1}_{\pm} \tag{4.5}$$

Here $\mathfrak{s}$ lies in the equatorial sphere while θ is the distance south of the north pole measured in degrees. The rectangular coordinates of $\omega = (x_1, x_2, ..., x_n)$ are uniquely recovered from $\mathfrak{s}$ and θ by the rule

$$\omega = (\cos\theta\,,\ \mathfrak{s}\cdot\sin\theta)\,, \qquad 0 < \theta < \pi \tag{4.6}$$

The $(n-1)$-surface area of $\mathbb{S}^{n-1}$ is expressed in terms of the meridian angle as

$$\mathrm{d}\omega \ = \ (\sin\theta)^{n-2}\ \mathrm{d}\theta\, \mathrm{d}\mathfrak{s}\ , \qquad \int_{\mathbb{S}^{n-1}} \mathrm{d}\omega \ = \ \omega_{n-1}. \tag{4.7}$$

where $\mathrm{d}\mathfrak{s}$ stands for the $(n-2)$-surface area of the equatorial sphere. Therefore,

$$\int_0^\pi \sin^{n-2}\theta\ \mathrm{d}\theta = \frac{\omega_{n-1}}{\omega_{n-2}} \tag{4.8}$$

We turn now to the basic examples of mappings between annuli.

4. Radial stretching

These are the mappings which change the radial distance but leave the spherical coordinate fixed. Let $\mathscr{AC}\,[r, R]$, $0 \leqslant r \leqslant R < \infty$, denote the space of absolutely continuous functions on a closed interval $[r, R]$. Associated to every $H \in \mathcal{AC}\,[r, R]$ is a mapping $h : \mathbb{A} \to \mathbb{A}^*$ of the annulus $\mathbb{A} = \{x\ ;\quad r \leqslant |x| \leqslant R\}$ onto an annulus $\mathbb{A}^* = \{y\ ;\quad r_* \leqslant |y| \leqslant R_*\}$, defined by the rule

$$h(x) = H(r)\,\omega\,, \qquad \text{where } \ r = |x| \ \text{ and } \omega = \frac{x}{|x|} \in \mathbb{S}^{n-1} \tag{4.9}$$

Here we have

$$r_* = \min_{r \leqslant t \leqslant R} |H(t)| \ \leqslant \ \max_{r \leqslant t \leqslant R} |H(t)| = R_* \tag{4.10}$$

We call h a radial stretching and refer to $H : [r, R] \to [r_*, R_*]$ as its normal strain. Note that a composition of two such mappings is again a radial stretching; absolute continuity is not lost.

5. Spherical mappings

A generalization of radial stretching is the so-called *spherical mapping*

$$h(x) = H(|x|)\,\Phi(\tfrac{x}{|x|})\ , \qquad \text{where} \qquad \Phi : \mathbb{S}^{n-1} \to \mathbb{S}^{n-1} \tag{4.11}$$

Here, as in the case of radial stretchings, the normal strain function is assumed to be absolutely continuous. The spherical part $\Phi : \mathbb{S}^{n-1} \to \mathbb{S}^{n-1}$, called *tangential tension,* is continuous and weakly differentiable. In what follows we will be actually concerned with homeomorphisms $\Phi : \mathbb{S}^{n-1} \to \mathbb{S}^{n-1}$ in the Sobolev class

$\mathscr{W}^{1,n}(\mathbb{S}^{n-1}, \mathbb{S}^{n-1})$. Composition of two spherical mappings results in the composition of their normal strain functions and tangential tensions, respectively.

CHAPTER 5

Radial n-Harmonics

In this chapter we identify all solutions to the n-harmonic equation

$$\text{div} \, \| Dh \|^{n-2} Dh = 0 \;\; \text{of the form} \;\; h(x) = H(|x|) \, \frac{x}{|x|} \tag{5.1}$$

Such solutions, called *radial n-harmonics*, will be originally defined in the annulus

$$\mathbb{A} = \mathbb{A}(a,b) = \{x \, ; \quad a < |x| < b\} \, , \qquad 0 < a < b < \infty \tag{5.2}$$

Concerning regularity, we assume that the *strain function* H belongs to the Sobolev space

$$\mathscr{W}^{1,n}_{\text{loc}}(a,b) \subset \mathscr{C}^{\alpha}_{\text{loc}}(a,b) \, , \qquad \alpha = \frac{n-1}{n} \tag{5.3}$$

We are going to give a clear account of how to solve the n-harmonic equation for the radial n-harmonics. But before, let us state in advance that the radial n-harmonics will actually extend as n-harmonics to the entire punctured space $\mathbb{R}^n_\circ$. Even more, they will extend continuously to the Möbius space $\hat{\mathbb{R}}^n = \mathbb{R}^n \cup \{\infty\}$, one point compactification of $\mathbb{R}^n$.

Let $\mathcal{H}_n$ denote the class of all n-harmonics in $\hat{\mathbb{R}}^n$, the following elementary transformations of variables x and h preserve this class.

- *Rescaling*;

$$h(x) \in \mathcal{H}_n \text{ implies } \lambda \, h(kx) \in \mathcal{H}_n \text{ for every } k > 0 \text{ and } \lambda \in \mathbb{R}$$

- *Reflection*;

$$h(x) \in \mathcal{H}_n \text{ implies } h\left(x|x|^{-2}\right) \in \mathcal{H}_n$$

The radial n-harmonics in $\mathbb{A} = \{x \, ; \;\; a < |x| < b\}$, $0 < a < b < \infty$ are none other than the unique solutions of the Dirichlet problem

$$\begin{cases} \text{div} \, \| Dh \|^{n-2} Dh = 0 & \text{for } a < |x| < b \\ h(x) = \alpha x & \text{for } |x| = a \\ h(x) = \beta x & \text{for } |x| = b \end{cases} \tag{5.4}$$

where α and β can be any real numbers, so $H(a) = \alpha \, a$ and $H(b) = \beta \, b$. It is understood here that h extends continuously to the closed annulus $\mathbb{A}\,[a,b]$. We shall distinguish four so-called *principal n-harmonics* in $\hat{\mathbb{R}}^n$ and use them to generate all radial n-harmonics via rescaling of the variables x and h. For this we need some computation.

1. The n-Laplacian for the strain function

Suppose $H \in \mathscr{C}^2(a,b)$, $0 \leqslant a < b \leqslant \infty$. Thus the derivatives of the radial stretching

$$h(x) = H(|x|)\,\frac{x}{|x|} \tag{5.5}$$

exist up to order 2. The differential matrix of h can be computed as

$$Dh(x) = \frac{H(|x|)}{|x|}\mathbf{I} + \frac{|x|\dot{H}(|x|) - H(|x|)}{|x|} \cdot \frac{x \otimes x}{|x|^2} \tag{5.6}$$

Hereafter $\dot{H}$ stands for the derivative of $H = H(t)$, $a < t < b$. Hence the Hilbert-Schmidt norm of the differential matrix is expressed by means of H as

$$\| \, Dh(x) \, \|^{\,2} = \operatorname{Tr}(D^*h \cdot Dh) = \dot{H}^2 + (n-1)\frac{H^2}{|x|^2} \tag{5.7}$$

A lengthy computation leads to the following explicit formula for the n-Laplacian

$$\begin{aligned} \operatorname{div} \| \, Dh \, \|^{\,n-2} Dh \;\; &= \;\; \frac{(n-1)|Dh|^{n-4}}{|x|^4} \cdot \Big\{ t^2(t^2\dot{H}^2 + H^2)\ddot{H} + \\ +t^3\dot{H}^3 \;\; &+ \;\; (n-3)t^2\dot{H}^2 H + t\dot{H}H^2 - (n-1)H^3 \Big\} \frac{x}{|x|} \end{aligned} \tag{5.8}$$

where $t = |x|$. Our subsequent analysis will be based on the *elasticity function* already mentioned at (1.30),

$$\eta = \eta_H = \frac{t\dot{H}(t)}{H(t)}. \tag{5.9}$$

Let us make certain here that we shall never deal with the case in which both $\dot{H}$ and H vanish simultaneously, so η will be well defined even when H vanishes. Now h is n-harmonic if and only if

$$\left(1+\eta^2\right) t^2 \ddot{H} = (1-\eta)\left[\eta^2 + (n-2)\eta + n - 1\right] \cdot H \tag{5.10}$$

where the case $\eta = \infty$ should read as $H = 0$. In dimension $n = 2$ the n-harmonic equation reduces to the familiar linear Cauchy-Euler equation for H

$$t^2\ddot{H} + t\dot{H} - H = 0 \tag{5.11}$$

It has two linearly independent solutions, denoted by

$$H_\circ(t) = t \quad \text{and} \quad H_\infty(t) = \frac{1}{t} \tag{5.12}$$

They generate all solutions by means of linear combinations. We pick out two particular solutions

$$H_+(t) = \frac{1}{2}\left(t + \frac{1}{t}\right) \quad \text{and} \quad H_-(t) = \frac{1}{2}\left(t - \frac{1}{t}\right) \tag{5.13}$$

and refer to $H_\circ, H_\infty, H_+$ and H_- as principal solutions. The reason for preferring these principal solutions to the ones in (5.12) is simply that we will be able to generate all radial n-harmonics, simply be rescaling the principal ones. Such a set of principal solutions will be particularly useful in higher dimensions where the n-Laplacian is nonlinear. For $n = 2$ the corresponding complex principal harmonics are

$$h_\circ(z) = z\,, \qquad h_\infty(z) = \frac{1}{z}\,, \tag{5.14}$$

$$h_+(z) = \frac{1}{2}\left(z + \frac{1}{\bar{z}}\right) \quad \text{and} \quad h_-(z) = \frac{1}{2}\left(z - \frac{1}{\bar{z}}\right) \tag{5.15}$$

The reader may wish to observe that $h_\circ$ and h_∞ map $\hat{\mathbb{R}}^2$ univalently onto itself, whereas $h_+, h_- : \hat{\mathbb{R}}^2 \to \hat{\mathbb{R}}^2$ have a branch set (folding) along the unit circle. Precisely, h_+ covers the exterior of the unit disk twice while h_- is a double cover of the entire space $\hat{\mathbb{R}}^2$. In spite of nonlinearity the analogous ordinary differential equations in higher dimensions are still effectively solvable, with some computational efforts. The key is the following identity which follows from (5.8).

LEMMA 5.1. *Let $h(x) = H(|x|)\frac{x}{|x|}$ be a radial stretching of class $\mathscr{C}^2(\mathbb{A}, \mathbb{R}^n)$, where $\mathbb{A} = \{x\,;\;\; a < |x| < b\}$. Then, with the notation $|x| = t$, we have*

$$\begin{aligned} &\frac{n\,\dot{H}}{n-1}\operatorname{div} \|\, Dh\,\|^{\,n-2} Dh \;= \\ &-\frac{x}{t^{n+1}}\frac{d}{dt}\left\{\left[(n-1)H^2 + t^2\dot{H}^2\right]^{\frac{n-2}{2}}\left(H^2 - t^2\dot{H}^2\right)\right\} \end{aligned} \tag{5.16}$$

2. The principal solutions

The main idea is to eliminate one derivative in the second order n-harmonic equation. This will lead us to a problem of solving a first order ODE with one free parameter. For nonlinear equations such a procedure is rather tricky. Fortunately, there is a very satisfactory realization of this idea due to the identity (5.16). We wish to express the n-harmonic equation for $h(x) = H(|x|)\frac{x}{|x|}$ in the form

$$\mathcal{L}H \overset{\text{def}}{=\!=} F(t, H, \dot{H}) = const. \tag{5.17}$$

By virtue of Lemma 5.1, the nonlinear differential operator $\mathcal{L}$ takes the following explicit form

$$\mathcal{L}H = \left[H^2 + \frac{t^2\dot{H}^2}{n-1}\right]^{\frac{n-2}{2}}\left(H^2 - t^2\dot{H}^2\right) = const. \;[1] \tag{5.18}$$

Certainly, all radial n-harmonics can be found by solving the so-called *characteristic equation*

$$\mathcal{L}H = const. \tag{5.19}$$

However, for the converse statement, caution must be exercised because this equation only implies that $\dot{H}\operatorname{div} \|\, Dh\,\|^{\,n-2}Dh = 0$. For example, the constant function $H(t) \equiv 1$ satisfies $\mathcal{L}H \equiv 1$, but the corresponding radial mapping $h(x) = \frac{x}{|x|}$ is not n-harmonic. [2] More generally, there are many more solutions to the equation $\mathcal{L}H \equiv 1$ which are constant on a subinterval. For example, one can paste $\frac{x}{|x|}$ with a

---

[1] In general, guessing the identity such as (5.18) takes some efforts; though verifying it poses no challenge.

[2] This noninjective solution will play an important role in the sequel.

suitable n-harmonic map on the rest of the interval. Such solutions, however, cannot be $\mathscr{C}^2$-smooth; $\mathscr{C}^{1,1}$-regular at best. They, nevertheless solve the generalized n-harmonic equation (3.6).

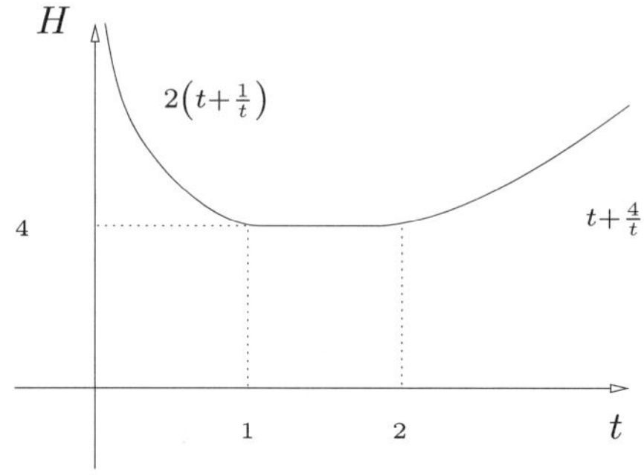

FIGURE 5.1. A solution which is constant on an interval.

Solutions $H = H(t)$ to the characteristic equation $\mathcal{L}H \equiv c$, that are constant in a proper subinterval like in the above figure, cannot be even $\mathscr{C}^{1,1}$-regular if $c \leqslant 0$.

Let us summarize our findings as:

DEFINITION 5.2. The term principal solution pertains to each of the following four functions of class $\mathscr{C}^2(0,\infty)$ which solve the equation $\mathcal{L}H = $ *constant*;

$$H_\circ(t) = t\,, \qquad \mathcal{L}H_\circ \equiv 0 \tag{5.20}$$

$$H_\infty(t) = \frac{1}{t}\,, \qquad \mathcal{L}H_\infty \equiv 0 \tag{5.21}$$

$$\mathcal{L}H_+ \equiv 1\,, \qquad H_+(1) = 1 \tag{5.22}$$

$$\mathcal{L}H_- \equiv -1\,, \qquad H_-(1) = 0 \tag{5.23}$$

We shall solve these Cauchy problems and show that both H_+ and H_- are actually $\mathscr{C}^\infty$-smooth.

The name "principal solutions" is given to $H_\circ$, H_∞, H_+ and H_- because they generate all radial n-harmonics. We denote the corresponding radial mapping by $h_\circ$, h_∞, h_+ and h_-, and call them *principal n-harmonics.* Precisely, we have

PROPOSITION 5.3. *Every radial n-harmonic mapping in the annulus $\mathbb{A} = \{x\,;\ a < |x| < b\}$ takes the form*

$$g(x) = \lambda \cdot h(kx)\,, \qquad \lambda \in \mathbb{R}\,,\ \ k > 0 \tag{5.24}$$

where $h \in \mathscr{C}^\infty(\mathbb{R}^n_\circ)$, is one of the four principal n-harmonics.

It should be observed, as a corollary, that radial n-harmonics are $\mathscr{C}^\infty$-smooth in the entire space $\mathbb{R}^n_\circ$.

3. The elasticity function

We shall distinguish four classes of the radial n-harmonics $h(x) = H(|x|)\frac{x}{|x|}$. The concept of so-called conformal elasticity underlines this distinction. In what follows it will never be the case that both H and $\dot{H}$ vanish simultaneously. Consider

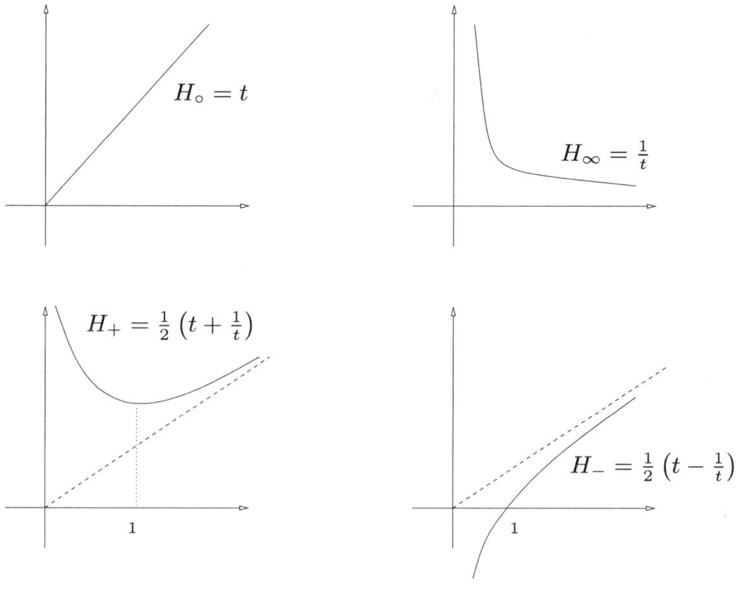

FIGURE 5.2. The principal solutions for $n = 2$. Similar ones for $n \geqslant 3$ are illustrated in Figures 5.7 and 5.10.

any function $H \in \mathscr{C}^1(r, R)$, $0 < r < R < \infty$, such that $H^2 + \dot{H}^2 > 0$. The elasticity function

$$\eta(t) = \eta_{_H}(t) = \frac{t\dot{H}(t)}{H(t)} \tag{5.25}$$

is continuous on (r, R). In general, it assumes values in the extended real line $\hat{\mathbb{R}} = \mathbb{R} \cup \{-\infty, \infty\}$. We recall that $h = H\big(|x|\big)\frac{x}{|x|}$ is:

- *Conformally contracting*, if $|\eta(t)| < 1$
- *Conformally expanding*, if $|\eta(t)| > 1$
- *Conformal*, if $|\eta(t)| = 1$

Observe that the equation $\eta(t) \equiv 1$ yields $H(t) = \lambda \cdot t$, while $\eta(t) = -1$ gives $H(t) = \frac{\lambda}{t}$. These are special cases of the power functions $H(t) = \lambda t^\alpha$, having constant elasticity $\eta(t) \equiv \alpha$. The above three classes are invariant under rescaling and inversion. Precisely, we have the formulas:

$$\begin{cases} \eta_{_H}(kt) = \eta_{_F}(t) & \text{where } F(t) = H(kt) \\ \eta_{_H}(1/t) = -\eta_{_F}(t) & \text{where } F(t) = H(1/t) \end{cases} \tag{5.26}$$

The elasticity function tells us something about the infinitesimal relative rate of change of the modulus of an annulus under the deformation $h = H\big(|x|\big)\frac{x}{|x|}$. Precisely, we have for $r < t < R$

$$\eta_{_H}(t) = \lim_{\epsilon \to 0} \frac{\operatorname{Mod} A^*(t+\epsilon) - \operatorname{Mod} A^*(t)}{\operatorname{Mod} A(t+\epsilon) - \operatorname{Mod} A(t)} \tag{5.27}$$

where $A(t) = \mathbb{A}\,(r,t)$ and $A^*(t) = h[A(t)]$.

We shall be concerned with $\mathscr{C}^1$-solutions to the characteristic equation

$$\mathcal{L}H = \left[H^2 + \frac{t^2 \dot{H}^2}{n-1} \right]^{\frac{n-2}{2}} \left(H^2 - t^2 \dot{H}^2 \right) \equiv c \tag{5.28}$$

This equation is invariant under rescaling and inversion of the independent variable t. Accordingly, all $\mathscr{C}^1$-solutions fall into three categories:

- H is contracting if $c > 0$, equivalently $|\eta(t)| < 1$
- H is expanding if $c < 0$, equivalently $|\eta(t)| > 1$
- H is conformal if $c = 0$, equivalently $|\eta(t)| = 1$,
  Furthermore, in the conformal case

$$\begin{cases} h \text{ is preserving the order of boundary components if} & \eta(t) \equiv 1\,,\ H(t) = \lambda t \\ h \text{ is reversing the order of boundary components if} & \eta(t) \equiv -1, H(t) = \frac{\lambda}{t} \end{cases}$$

This is easily seen by writing (5.28) as

$$\left(1 + \frac{\eta^2}{n-1}\right)^{\frac{n-2}{2}} (1 - \eta^2) = \frac{c}{|H|^n} \tag{5.29}$$

We then distinguish four classes of the radial n-harmonics:

$$\mathcal{H}_n = \mathcal{H}_+ \cup \mathcal{H}_1 \cup \mathcal{H}_\circ \cup \mathcal{H}_\infty \tag{5.30}$$

where

$$\begin{aligned} \mathcal{H}_+ &= \left\{ h = H(|x|)\, \frac{x}{|x|}\,; \quad \mathcal{L}H \equiv c > 0 \right\} \\ \mathcal{H}_- &= \left\{ h = H(|x|)\, \frac{x}{|x|}\,; \quad \mathcal{L}H \equiv c < 0 \right\} \\ \mathcal{H}_\circ &= \left\{ h\,; \quad h(x) = \lambda x\,, \quad \lambda \in \mathbb{R}_\circ \right\} \\ \mathcal{H}_\infty &= \left\{ h\,; \quad h(x) = \frac{\lambda \cdot x}{|x|^2}\,, \quad \lambda \in \mathbb{R}_\circ \right\} \end{aligned}$$

We strongly emphasize again that not every solution to the equation $\mathcal{L}H \equiv c$ defines a radial n-harmonic map, due to lack of regularity. There are Lipschitz solutions to $\mathcal{L}H = 0$ lacking $\mathscr{C}^1$-regularity

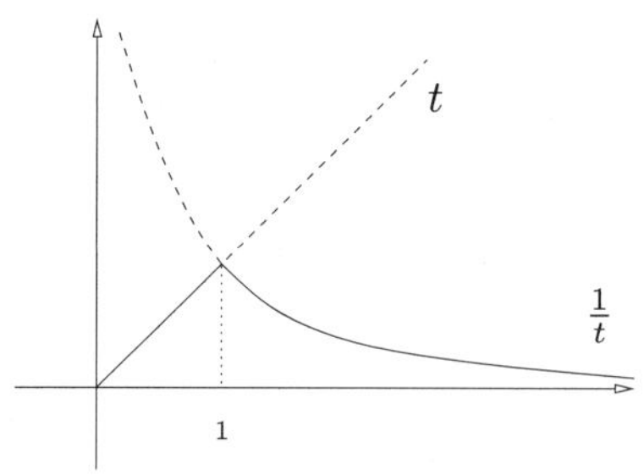

FIGURE 5.3. Lipschitz solution to $\mathcal{L}H = 0$.

Such solutions are not particularly desirable. Even $\mathscr{C}^1$-solutions to the equations $\mathcal{L}H = 1$ need not give radial n-harmonics. For example, if we paste $H_+(t) =$

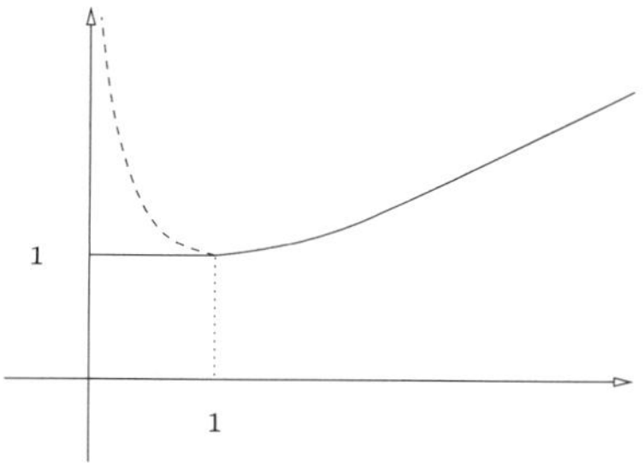

FIGURE 5.4. A $\mathscr{C}^1$-solution to $\mathcal{L}H = 1$.

$\frac{1}{2}\left(t + \frac{1}{t}\right)$ $1 \leqslant t < \infty$ with function identically equal to 1 for $0 < t \leqslant 1$ then the resulting function will become a $\mathscr{C}^{1,1}$-solution to the equation $\mathcal{L}H = 1$ on $(0, \infty)$.

The radial stretching obtained in this way fails to be n-harmonic on the punctured disk $\mathbb{A}(0,1)$. Thus we see various degenerations of $\mathscr{C}^1$-solutions, suggesting that we should restrict ourselves to the solutions of class $\mathscr{C}^2(0,\infty)$.[3]

The two principal solutions H_+ and H_- are particularly interesting and important to the forthcoming results, so we devote next two sections for a detailed treatment.

4. The principal solution H_+ (conformal contraction)

We find H_+ as a $\mathscr{C}^2$-solution to the Cauchy problem

$$\mathcal{L}H = \left[H^2 + \frac{t^2\dot{H}^2}{n-1}\right]^{\frac{n-2}{2}} \left(H^2 - t^2\dot{H}^2\right) \equiv 1\,, \qquad H(1) = 1 \tag{5.31}$$

The general idea behind our method of solving this equation is as follows. Differentiating (5.31) yields a first order equation for the elasticity function,

$$\frac{(1+\eta^2)\,\dot{\eta}}{(1-\eta^2)\,(\eta^2+n-1)} = \frac{1}{t}\,, \qquad |\eta(t)| < 1 \tag{5.32}$$

This can also be directly derived from (5.8). With the equation (5.32) at hand we now proceed to the explicit computation. Consider the function $\Gamma_+ = \Gamma_+(s)$ defined for $-1 < s < 1$ by the rule

$$\begin{aligned} \Gamma_+(s) &= \exp \int_0^s \frac{(1+\tau^2)\,d\tau}{(1-\tau^2)(\tau^2+n-1)} \\ &= \sqrt[n]{\tfrac{1+s}{1-s}} \exp\left[\tfrac{2-n}{n\sqrt{n-1}} \tan^{-1} \tfrac{s}{\sqrt{n-1}}\right] \end{aligned} \tag{5.33}$$

Obviously Γ_+ is strictly increasing from zero to infinity, thus invertible.

Note the identities

$$\begin{aligned} \Gamma_+(0) &= 1 \\ \dot{\Gamma}_+(0) &= \tfrac{1}{n-1} \\ \Gamma_+(-s)\,\Gamma_+(s) &= 1 \end{aligned} \tag{5.34}$$

We then examine the inverse function defined for $0 < t < \infty$,

$$u(t) = u_+(t) = \Gamma_+^{-1}(t)\,, \quad u(1) = 0\,, \quad \dot{u}(1) = n-1 \tag{5.35}$$

[3]In the expanding case $\mathscr{C}^1$ solutions are always smooth.

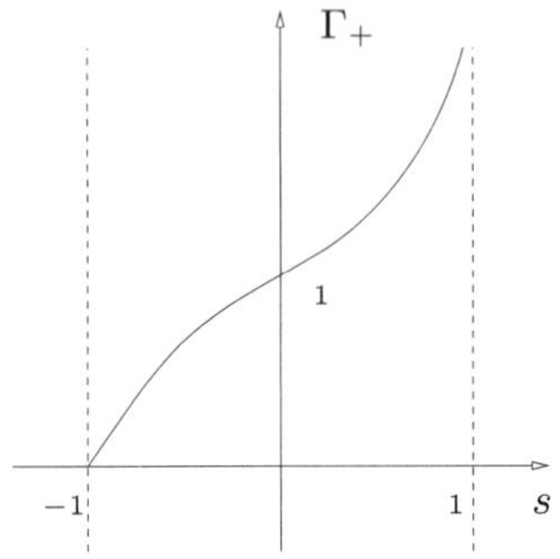

FIGURE 5.5. Γ_+ is strictly increasing and $\mathscr{C}^\infty$-smooth.

The identity $\Gamma_+(-s)\,\Gamma_+(s) \equiv 1$ translates as the antisymmetry rule for u

$$u\left(\frac{1}{t}\right) = -u(t)\,, \qquad t > 0 \tag{5.36}$$

Implicit differentiation of the equation $\Gamma_+\big(u(t)\big) \equiv t$ yields

$$\dot u\,\dot\Gamma_+(u) = 1\,, \qquad\qquad t\,\dot u(t) = \frac{\Gamma_+(u)}{\dot\Gamma_+(u)} \tag{5.37}$$

We shall now introduce $H_+ = H_+(t)$, $0 < t < \infty$, and prove that H_+ is the principal solution, namely

$$H_+(t) = \left(1 + \frac{u^2}{n-1}\right)^{\frac{1}{n}-\frac{1}{2}} \left(1-u^2\right)^{-\frac{1}{n}} \geqslant 1 \tag{5.38}$$

From this formula it follows that u is the elasticity function of H_+. Indeed,

$$\eta = \eta_+(t) = \frac{t\dot H_+}{H_+} = \frac{u \cdot t\,\dot u\,(1+u^2)}{(1-u^2)(n-1+u^2)} = u \cdot t\dot u\,\frac{\dot\Gamma_+(u)}{\Gamma_+(u)} = u(t) \tag{5.39}$$

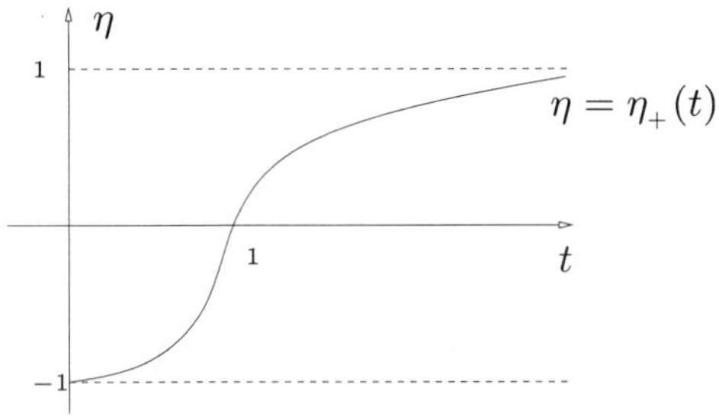

FIGURE 5.6. The elasticity function of H_+.

Now, the equation $\mathcal{L}H_+ \equiv 1$ can easily be verified

$$\begin{aligned} \mathcal{L}H_+ &= |H_+|^n \left[1 + \frac{\eta^2}{n-1}\right]^{\frac{n-2}{2}} (1-\eta^2) = \\ & \quad |H_+|^n \left[1 + \frac{u^2}{n-1}\right]^{\frac{n-2}{2}} (1-u^2) \equiv 1 \end{aligned} \tag{5.40}$$

This last step is immediate from (5.38). It is worth noting that $H_+ \in \mathscr{C}^\infty(0,\infty)$ and $\dot{H}_+(1) = 0$. Antisymmetry rule at (5.36) results in a symmetry rule for H_+

$$H_+\left(\frac{1}{t}\right) = H_+(t) \tag{5.41}$$

Furthermore, it follows from (5.10) that $\ddot{H}_+(t) > 0$. Thus H_+ is strictly convex, see Figure 5.7.

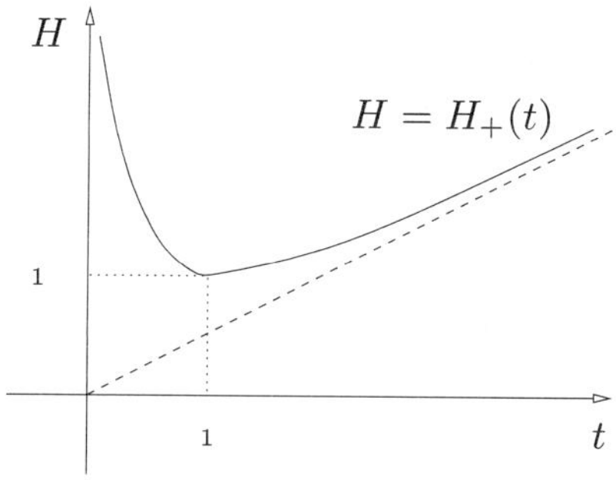

FIGURE 5.7. The principal solution H_+.

Let us take a look at the behavior of H_+ near zero and infinity. It has an asymptote at infinity with the slope

$$\Theta = \Theta_+ = \left(1 - \frac{1}{n}\right)^{\frac{n-2}{2n}} 4^{-\frac{1}{n}} \exp\left[\frac{n-2}{n\sqrt{n-1}} \tan^{-1}\frac{1}{\sqrt{n-1}}\right] \tag{5.42}$$

Indeed, under the notation above, we have the following equation for $t = \Gamma(s)$

$$H^n(t) - \Theta^n t^n = \frac{1}{1-s^2}\left(1 + \frac{s^2}{n-1}\right)^{1-\frac{n}{2}} - \Theta^n \Gamma_+^n(s) \stackrel{\text{def}}{=\!=} \frac{A(s)}{s-1}$$

Note that $A = A(s)$ is $\mathscr{C}^\infty$-smooth on $(-1,\infty)$. It vanishes at $s = 1$. Application of L'Hospital rule shows that

$$\lim_{t\to\infty}\left[H^n(t) - \Theta^n t^n\right] = A'(1) > 0\,, \tag{5.43}$$

whence the inequality

$$0 < H(t) - \Theta\, t \leqslant \frac{C}{t^{n-1}}\,, \qquad C = \frac{A'(1)}{n\Theta^{n-1}} > 0 \tag{5.44}$$

In particular by (5.41) we conclude that

$$\lim_{t\to\infty}\frac{H(t)}{t} = \Theta = \lim_{t\to 0} t\, H(t) \tag{5.45}$$

5. The principal solution H_- (conformal expansion)

The principal solution $H = H_- \in \mathscr{C}^\infty(0,\infty)$ is obtained by solving the following Cauchy problem for the characteristic equation

$$\mathcal{L}H = \left[H^2 + \frac{t^2\dot{H}^2}{n-1}\right]^{\frac{n-2}{2}}\left(H^2 - t^2\dot{H}^2\right) \equiv -1\,, \quad H(1) = 0 \tag{5.46}$$

Obviously $\dot{H}$ cannot vanish. For an explicit computation we introduce the following function

$$\Gamma_-(s) = \sqrt{\frac{1+s}{1-s}} \exp\left[\frac{n-2}{n\sqrt{n-1}} \tan^{-1}\left(s\sqrt{n-1}\right)\right] \tag{5.47}$$

where $-1 < s < 1$. Its logarithmic derivative is computed as

$$\frac{\dot{\Gamma}_-(s)}{\Gamma_-(s)} = \frac{1+s^2}{(1-s^2)\,[1+(n-1)s^2]} > 0 \tag{5.48}$$

This shows that $\Gamma_- : (-1,1) \to (0,\infty)$ is strictly increasing. Consequently, we consider its inverse function

$$u(t) = u_-(t) = \Gamma^{-1}(t)\,, \qquad 0 < t < \infty \tag{5.49}$$

The graphs of Γ_- and u_- are very similar to those of Γ_+ and u_+, though they are not the same. However, this time the same identity $\Gamma(-s)\,\Gamma(s) = 1$ yields different transformation rule for u,

$$u\left(\frac{1}{t}\right) = -u(t)\,, \qquad 0 < t < \infty \tag{5.50}$$

It is at this point that the analogy with H_+ ends. We define the principal solution H_- as

$$H(t) = H_-(t) = u\left(u^2 + \frac{1}{n-1}\right)^{\frac{1}{n}-\frac{1}{2}} \left(1-u^2\right)^{-\frac{1}{n}} \tag{5.51}$$

Note that H_- changes sign at $t = 1$, exactly where u vanishes. Basic properties of $H_-(t)$ can be derived directly from (5.51). Let us state some of them. We observe the first difference between H_+ and H_- in the antisymmetry formula

$$H_-\left(\frac{1}{t}\right) = -H_-(t) \tag{5.52}$$

which is immediate from (5.50). Next for $t \neq 1$ we compute the elasticity function for H_-,

$$\begin{aligned} \eta = \eta_-(t) &= \frac{t\,\dot{H}_-(t)}{H_-(t)} = \frac{(1+u^2)\,t\dot{u}}{u\,(1-u^2)\,[1+(n-1)u^2]} \\ &= \frac{t\,\dot{u}}{u} \cdot \frac{\dot{\Gamma}_-(u)}{\Gamma_-(u)} = \frac{1}{u} \end{aligned} \tag{5.53}$$

Now the characteristic equation for $H = H_-(t)$ is easily verified

$$\begin{aligned} \mathcal{L}H_- &= |H_-|^n \left[1 + \frac{\eta^2}{n-1}\right]^{\frac{n-2}{2}} (1-\eta^2) \\ &= \left|\frac{H_-}{u}\right|^n (u^2-1)\left[u^2 + \frac{1}{n-1}\right]^{\frac{n-2}{2}} \equiv -1 \end{aligned} \tag{5.54}$$

Our analysis of convexity properties of H_- is based on the equation (5.10), which we write as

$$(1+u^2)t^2\,\ddot{H} = (u-1)\left[(n-1)u^2 + (n-2)u + 1\right]\frac{H}{u} \tag{5.55}$$

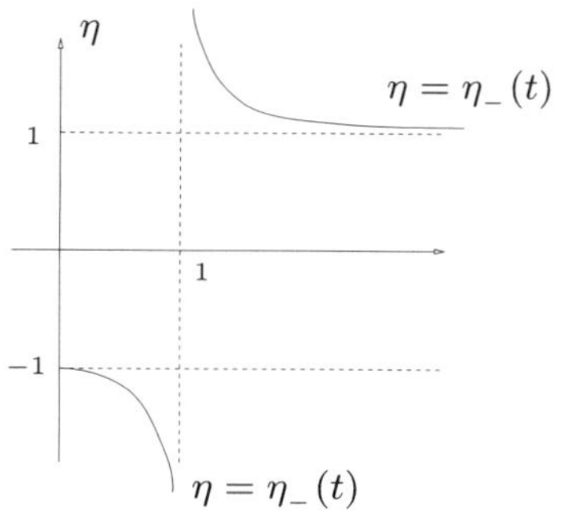

FIGURE 5.8. The graph of the elasticity function for H_-.

for all $t \neq 1$. Since the quotient $\frac{H}{u}$ is positive we see that

$$\operatorname{sgn} \ddot{H}_- = -\operatorname{sgn}\left[(n-1)u^2 + (n-2)u + 1\right] \tag{5.56}$$

The key observation is that for $n = 2, 3, 4, 5, 6$ the polynomial $P = P_n(u) = (n-1)u^2 + (n-2)u + 1$ has no roots, meaning that H_- is concave. Rather unexpectedly, H_- is no longer concave when $n \geqslant 7$, because P_n has two roots in the interval $u \in (-1, 0)$.

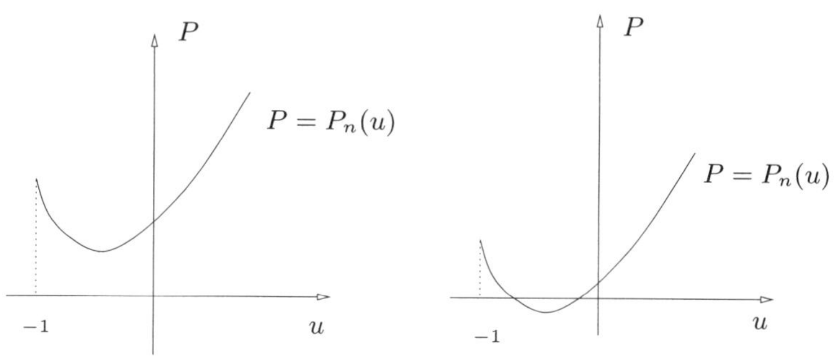

FIGURE 5.9. P_n is positive if $2 \leqslant n \leqslant 6$ and has two roots if $n \geqslant 7$.

As a consequence, the graphs of H_- exhibit two inflection points when $n \geqslant 7$.

Analysis of the asymptotic behavior of H_- is much the same as that for H_+. We only state the results. The slope of the asymptote at infinity equals

$$\Theta = \Theta_- = \left(1 - \frac{1}{n}\right)^{\frac{n-2}{2n}} 4^{-\frac{1}{n}} \exp\left[\frac{2-n}{n\sqrt{n-1}} \tan^{-1}\sqrt{n-1}\right] \tag{5.57}$$

and we have

$$0 < \Theta t - H_-(t) \leqslant \frac{C}{t^{n-1}} \tag{5.58}$$

In particular, for $H = H_-(t)$ it holds

$$\lim_{t\to\infty} \frac{H(t)}{t} = \Theta = -\lim_{t\to 0} t\, H(t) \tag{5.59}$$

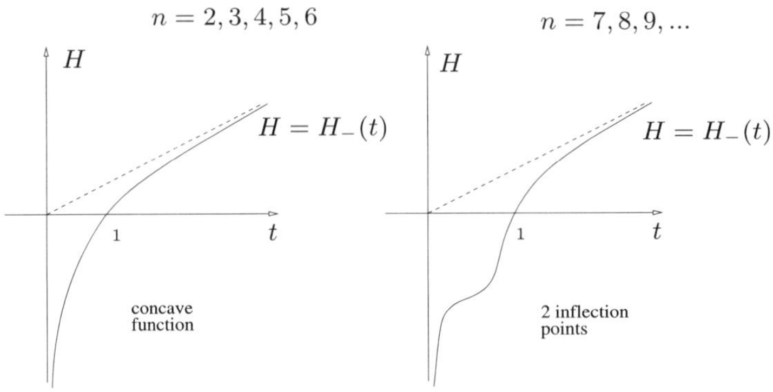

FIGURE 5.10. The principal solution H_-.

6. The boundary value problem for radial n-harmonics

Given $\alpha, \beta \in \mathbb{R}$ one may wish to look for functions $H \in \mathscr{C}^2(a,b) \cap \mathscr{C}[a,b]$, $0 < a < b < \infty$, which satisfy the conditions

$$\begin{cases} \mathcal{L}H \equiv c \\ H(a) = \alpha \\ H(b) = \beta \end{cases} \tag{5.60}$$

Here the constant c is also viewed as unknown, otherwise the system (5.60) would be overdetermined (ill-posed). Let us begin with the easy case $\beta = 0$. The solution is given by the formula

$$H(t) = \lambda H_-(kt)\,, \qquad k = \frac{1}{b} \qquad \lambda = \frac{\alpha}{H\left(\frac{a}{b}\right)} \tag{5.61}$$

From now on we assume that $\beta \neq 0$. There are five cases to consider.
Case 1. Suppose that

$$-\infty < \frac{\alpha}{\beta} < \frac{a}{b} \tag{5.62}$$

The solution will be found in the form $H(t) = \lambda H_-(kt)$. One has to show that there exist real number λ and a positive number k such that

$$\begin{cases} \lambda H_-(ka) = \alpha \\ \lambda H_-(kb) = \beta \end{cases} \tag{5.63}$$

We eliminate λ by dividing the equations

$$Q(k) = \frac{H_-(ka)}{H_-(kb)}\,, \qquad \text{for } k > \frac{1}{b} \tag{5.64}$$

Note that $\lim\limits_{k \to \frac{1}{b}} Q(k) = -\infty$. On the other hand using (5.59) we see that

$$\lim_{k\to\infty} Q(k) = \frac{a}{b}\left(\lim_{k\to\infty} \frac{\frac{H_-(ka)}{ka}}{\frac{H_-(kb)}{kb}}\right) = \frac{a\,\Theta_-}{b\,\Theta_-} = \frac{a}{b} \tag{5.65}$$

By Mean Value Theorem there exists $\infty > k_\circ > \frac{1}{b}$ such that

$$Q(k_\circ) = \frac{H_-(k_\circ a)}{H_-(k_\circ b)} = \frac{\alpha}{\beta} \tag{5.66}$$

Finally, conditions (5.63) are satisfied with $\lambda = \frac{\beta}{H_-(k_\circ b)}$.

The next case is obvious; a linear function is a solution.

Case 2. Suppose

$$\frac{\alpha}{\beta} = \frac{a}{b} \tag{5.67}$$

Then the solution is given by

$$H(t) = \tfrac{\alpha}{\beta}\, t = \alpha\, H_\circ\left(\tfrac{t}{a}\right) \tag{5.68}$$

Case 3. Suppose now that

$$\frac{a}{b} < \frac{\alpha}{\beta} < \frac{b}{a} \tag{5.69}$$

We are looking for the solution in the form $H(t) = \lambda H_+(kt)$. As before, one needs to find $k_\circ$ such that

$$\frac{H_+(k_\circ a)}{H_+(k_\circ b)} = \frac{\alpha}{\beta} \tag{5.70}$$

For this reason we consider the function

$$Q(k) = \frac{H_+(ka)}{H_+(kb)}, \qquad k > 0 \tag{5.71}$$

With the aid of the asymptotic formula (5.45) we see that $\lim_{k\to\infty} Q(k) = \frac{a}{b}$ and $\lim_{k\to 0} Q(k) = \frac{b}{a}$. By Mean Value Theorem the equation (5.70) holds for some $k_\circ > 0$. Then the parameter λ can be chosen to yield the boundary conditions at (5.63).

Next case is immediate.

Case 4. Suppose

$$\frac{\alpha}{\beta} = \frac{b}{a} \tag{5.72}$$

The solution is given by the formula $H(t) = \frac{\alpha a}{t} = \alpha\, H_\infty\left(\frac{t}{a}\right)$.

Case 5. Suppose finally that

$$\frac{\alpha}{\beta} > \frac{b}{a} \tag{5.73}$$

We are looking for the solution in the form $H(t) = \lambda H_-(kt)$, so we examine the function

$$Q(k) = \frac{H_-(ka)}{H_-(kb)}, \qquad \text{with } k < \frac{1}{b} \tag{5.74}$$

By virtue of formula (5.59) we see that $\lim_{k\to\frac{1}{b}} Q(k) = +\infty$ and $\lim_{k\to 0} Q(k) = \frac{b}{a}$. This case is completed by again invoking Mean Value Theorem; there is $k_\circ < \frac{1}{b}$ such that

$$\frac{H_-(k_\circ a)}{H_-(k_\circ b)} = \frac{\alpha}{\beta} \tag{5.75}$$

Then, we take λ to ensure that $\lambda H_-(k_\circ b) = \beta$.

In summing up this section, we note that the foregoing analysis, combined with the uniqueness of the Dirichlet problem for the n-harmonic equation, reveals that in fact the principal n-harmonics generate all radial n-harmonics. As an unexpected bonus we deduce that every radial n-harmonic map in a ring domain extends n-harmonically to the entire punctured space $\mathbb{R}^n_\circ = \mathbb{R}^n \setminus \{0\}$ and continuously to the Möbius space $\hat{\mathbb{R}}^n = \mathbb{R}^n \cup \{\infty\}$.

CHAPTER 6

Vector Calculus on Annuli

In this chapter we briefly review differential calculus on annuli.

1. Radial and spherical derivatives

In studying deformations of spherical rings one must consider the radial and spherical components of a differential map. Through every point $x \in \mathbb{R}^n_\circ$ there passes a sphere

$$\mathbb{S}^{n-1}_t = \{y \in \mathbb{R}^n ; \quad |y| = t\}, \qquad t = |x| \tag{6.1}$$

Its tangent hyperplane at x is given by

$$\mathbb{T}_x = \{\xi ; \quad \langle x, \xi \rangle = t^2\} \tag{6.2}$$

Fix an orthonormal basis for $\mathbb{T}_x$, say $\mathbf{T} = \{T_2, T_3, ..., T_n\}$.[1] Let $\Omega \subset \mathbb{R}^n_\circ$ be a domain and $h : \Omega \to \mathbb{R}^n$ a mapping having first order partial derivatives defined at $x \in \Omega$. The radial (or normal) derivative of h at x is a vector defined by the rule

$$h_{\mathbf{N}}(x) = Dh(x) \cdot \mathbf{N} = \frac{x_1 h_{x_1} + ... + x_n h_{x_n}}{|x|} \tag{6.3}$$

where $\mathbf{N} = \frac{x}{|x|}$ is called the radial (or normal) vector. In an exactly similar way we define the spherical (or tangential) derivatives

$$h_{\mathbf{T}}(x) = [h_{T_2}, ..., h_{T_n}] \ , \quad h_{T_\nu} = Dh(x) \cdot T_\nu \ , \quad \nu = 2, 3, ..., n \tag{6.4}$$

This is an $(n-1)$-tuple of vectors in $\mathbb{R}^n$, conveniently considered as column vectors of an $n \times (n-1)$-matrix. Continuing in this fashion we view the differential of Dh as an $n \times n$-matrix whose column vectors are $h_{\mathbf{N}}, h_{T_2}, ..., h_{T_n}$. Of course, such a matrix representation of Dh depends on choice of the orthonormal frame $\mathbf{N}, T_2, ..., T_n$

$$Dh(x) \approx \begin{bmatrix} | & | & & | \\ | & | & & | \\ h_{\mathbf{N}} & h_{T_2} & & h_{T_n} \\ | & | & & | \\ | & | & & | \end{bmatrix} \tag{6.5}$$

However, some differential expressions do not depend on the frame. For example, the Hilbert-Schmidt norm of the differential matrix,

$$\| Dh \|^2 = |h_{\mathbf{N}}|^2 + |h_{T_2}|^2 + ... + |h_{T_n}|^2 \tag{6.6}$$

Also, the spherical component of $\| Dh \|$, which we define by the formula

$$|h_{\mathbf{T}}| = \left(\frac{|h_{T_2}|^2 + ... + |h_{T_n}|^2}{n-1} \right)^{\frac{1}{2}} \geqslant \Big(|h_{T_2}| \cdots |h_{T_n}| \Big)^{\frac{1}{n-1}} \tag{6.7}$$

[1] It is not always possible to choose such a basis continuously depending on $x \in \mathbb{R}^n_\circ$

is frame free. Equality holds if and only if $|h_{T_2}| = ... = |h_{T_n}|$. Thus we have

$$\| Dh \|^2 = |h_{\mathbf{N}}|^2 + (n-1)\, |h_{\mathbf{T}}|^2 \tag{6.8}$$

Another differential quantity of interest to us is the Jacobian determinant

$$J(x,h) = \det Dh = \langle h_{\mathbf{N}} \,,\, h_{T_2} \times ... \times h_{T_n} \rangle \tag{6.9}$$

where the cross product of spherical derivatives is controlled by using the point-wise Hadamard's inequality.

$$|h_{T_2} \times ... \times h_{T_n}| \leqslant |h_{T_2}| \cdots |h_{T_n}| \tag{6.10}$$

Hence

$$J(x,h) \leqslant |h_{\mathbf{N}}| \cdot |h_{\mathbf{T}}|^{n-1} \tag{6.11}$$

Equality occurs if and only if the vectors $h_{\mathbf{N}}, h_{T_2}, ..., h_{T_n}$ are mutually orthogonal, positively oriented and $|h_{T_2}| = ... = |h_{T_n}|$. This amounts to saying that the Cauchy-Green tensor of h must be a diagonal matrix

$$\mathbf{C}(x,h) \overset{\text{def}}{=\!=} D^*h \cdot Dh = \begin{bmatrix} |h_N|^2 & 0 & \cdots & 0 \\ 0 & |h_T|^2 & \cdots & 0 \\ & & \ddots & \\ 0 & 0 & \cdots & |h_T|^2 \end{bmatrix} \tag{6.12}$$

The notion of elasticity that we have introduced for radial stretchings can be extended to all weakly differentiable mappings. Let $h : \mathbb{A} \to \mathbb{A}^*$ be any mapping of annuli, and $x_\circ \in \mathbb{A}$ be a point where h is differentiable.[2] We say that, at this point, h is

- *expanding*, if $|\, h_N(x_\circ) \,| > |\, h_T(x_\circ) \,|$, $\quad x_\circ$ is a point of expansion
- *contracting*, if $|\, h_N(x_\circ) \,| < |\, h_T(x_\circ) \,|$, $\quad x_\circ$ is a point of contraction

In what follows we shall be concerned only with the regular points, that is $\| Dh(x_\circ) \| \neq 0$. We define the *elasticity module* of h at the regular points by the rule

$$\eta = \eta_h(x) = \frac{|\, h_N(x) \,|}{|\, h_T(x) \,|} \in \hat{\mathbb{R}} \tag{6.13}$$

This is a measurable function taking values in the extended half line $[0, \infty]$. Note the following inequalities,

$$K_{\circ}(x,h) \;=\; \frac{|Dh(x)|^n}{J(x,h)} \geqslant \frac{|\, h_N \,|^n}{|\, h_N \,|\, |\, h_T \,|^{n-1}} = [\eta_h(x)]^{n-1} \tag{6.14}$$

$$\mathbb{K}_{\circ}(x,h) \;=\; \frac{\|\, Dh \,\|^n}{n^{\frac{n}{2}} J(x,f)} \geqslant \frac{\left(|\, h_N \,|^2 + (n-1)\,|\, h_T \,|^2\right)^{\frac{n}{2}}}{n^{\frac{n}{2}}\, |\, h_N \,|\, |\, h_T \,|^{n-1}}$$

$$= \frac{1}{\eta_h} \left(\frac{n-1}{n} + \frac{1}{n} \eta_h^2 \right)^{\frac{n}{2}} \tag{6.15}$$

We say that the deformation $h : \mathbb{A} \to \mathbb{A}^*$ at the given point $x_\circ$ is:

- *contracting*, if $\eta_h(x_\circ) < 1$
- *expanding*, if $\eta_h(x_\circ) > 1$

---

[2]Homeomorphisms of class $\mathscr{W}^{1,n}$ are differentiable at almost every point.

It will be clear later that the extremal deformations have the same elasticity type throughout the region. Furthermore, the elasticity module will not vanish inside the region. Thus for the extremal deformations the equation at (6.12) takes the form of a Beltrami system

$$D^*h \cdot Dh = J(x,h)^{\frac{2}{n}}\mathbf{K}(x) \tag{6.16}$$

where the distortion tensor $\mathbf{K}$ is a diagonal matrix of determinant 1

$$\mathbf{K} = \begin{bmatrix} \eta^{2-\frac{2}{n}} & 0 & \cdots & 0 \\ 0 & \eta^{-\frac{2}{n}} & \cdots & 0 \\ & & \ddots & \\ 0 & 0 & \cdots & \eta^{-\frac{2}{n}} \end{bmatrix} \tag{6.17}$$

For the radial stretching

$$h(x) = H(|x|)\frac{x}{|x|} \tag{6.18}$$

we find that

$$h_{\mathbf{N}} = \dot{H}(|x|)\cdot\mathbf{N}\,, \qquad h_{T_i} = \frac{H(|x|)}{|x|}\cdot T_i \tag{6.19}$$

Hence, the cross product of the spherical derivative is parallel to the radial vector.

$$h_{T_2}\times \ldots \times h_{T_n} = \left(\frac{H}{|x|}\right)^{n-1}\frac{x}{|x|} \tag{6.20}$$

The square of Hilbert-Schmidt norm of the differential is given by

$$\|\, Dh \,\|^{\,2} = \dot{H}^2 + (n-1)\frac{H^2}{t^2}\,, \qquad\qquad t = |x| \tag{6.21}$$

and the Jacobian determinant equals:

$$J(x,h) = \frac{\dot{H}\,H^{n-1}}{t^{n-1}} \tag{6.22}$$

2. Some differential forms

Integration of nonlinear differential expressions is best handled by using differential forms. Let us begin with the function $t = |x|$ and its differential

$$dt = \frac{x_1dx_1 + \ldots + x_ndx_n}{|x|} \tag{6.23}$$

The Hodge star of dt is an $(n-1)$-form.

$$\star\, dt = \frac{1}{|x|}\sum_{i=1}^{n}(-1)^i x_i\, dx_1\wedge\ldots\wedge dx_{i-1}\wedge dx_{i+1}\wedge\ldots\wedge dx_n \tag{6.24}$$

This form integrates naturally on spheres centered at the origin. We wish to normalize this form in such a way that the integrals will be independent of the sphere. In this way we come to what is known as standard area form on $\mathbb{S}^{n-1}$

$$\omega(x) = \frac{\star\,\mathrm{d}t}{t^{n-1}} = \sum_{i=1}^{n}(-1)^i\,\frac{x_1dx_1\wedge\ldots\wedge dx_{i-1}\wedge dx_{i+1}\wedge\ldots\wedge dx_n}{|x|^n} \tag{6.25}$$

Viewing ω as a differential form on punctured space $\mathbb{R}^n_\circ$, we find that $d\omega = 0$. By Stokes' theorem

$$\int_{|x|=t} \omega = \omega_{n-1}\,, \qquad \text{for all } t > 0 \tag{6.26}$$

Next, given any mapping $h = (h^1, ..., h^n) : \Omega \to \mathbb{R}^n_\circ$ of class $\mathscr{W}^{1,1}_{\mathrm{loc}}(\Omega, \mathbb{R}^n_\circ)$, we consider the pullback of ω via h

$$h^\sharp \omega = \sum_{i=1}^{n} (-1)^i \frac{h^i dh^1 \wedge ... \wedge dh^{i-1} \wedge dh^{i+1} \wedge ... \wedge dh^n}{|h|^n} \tag{6.27}$$

Under suitable regularity hypothesis, for instance if $h \in \mathscr{W}^{1,n-1}_{\mathrm{loc}}(\Omega, \mathbb{R}^n)$, and $|h(x)| \geqslant const > 0$, this form is also closed, meaning that $d\,(h^\star \omega) = h^\star\,(d\omega) = 0$. We view the exterior derivative of h as an n-tuple of 1-forms, $dh = \left[dh^1, ..., dh^n\right]$. With such a view $dh \wedge dt$ becomes an n-tuple of 2-forms. Further notation is self explanatory. For example,

$$|dh \wedge dt|^2 = \left|dh^1 \wedge dt\right|^2 + ... + |dh^n \wedge dt|^2 \tag{6.28}$$

Now the decomposition at (6.8) takes the form

$$\|\, Dh \,\|^2 = |dh|^2 = |dh \wedge \star dt|^2 + |dh \wedge dt|^2 \tag{6.29}$$

The reader may verify this formula by the following general algebraic identity $|\mathfrak{a}|^2 = |\mathfrak{a} \wedge \star N|^2 + |\mathfrak{a} \wedge N|^2$ for all covectors $\mathfrak{a} \in \bigwedge^1(\mathbb{R}^n)$. Just apply it to $\mathfrak{a}^i = dh^i$, $i = 1, 2, ..., n$ and $N = dt$, then add the resulting identities. A moment of reflection about (6.29) reveals that

$$dh \wedge \star dt = h_N\, dx \tag{6.30}$$

and

$$dh \wedge dt = \frac{1}{2|x|} \sum_{1 \leqslant i < j \leqslant n} \left(x_i h_{x_j} - x_j h_{x_i}\right) dx_i \wedge dx_j \tag{6.31}$$

Hence

$$|dh \wedge dt|^2 = (n-1)|h_T|^2 \tag{6.32}$$

PROPOSITION 6.1. *The following point-wise estimates hold*

$$|h_N| \geqslant |\, d|h| \wedge \star dt\,| \tag{6.33}$$

$$|h_T|^{n-1} \geqslant |h|^{n-1} \left|dt \wedge h^\sharp \omega\right| \tag{6.34}$$

PROOF. Formula (6.30) holds for scalar functions in place of h we apply it to $|h|$

$$|h_N| \geqslant |\,|h|_N\,| = |\, d|h| \wedge \star dt\,| \tag{6.35}$$

establishing the inequality (6.33).

The proof of the estimate (6.34) is more involved. Using (6.32), we are reduced to showing that

$$\left|\frac{dh \wedge dt}{n-1}\right|^{n-1} \geqslant \left|\sum_{i=1}^{n} (-1)^i \frac{h^i}{|h|} dh^1 \wedge ... \wedge dh^{i-1} \wedge dh^{i+1} \wedge ... \wedge dh^n \wedge dt\right| \tag{6.36}$$

By Schwarz inequality it suffices to prove that

$$\left|\frac{dh \wedge dt}{n-1}\right|^{n-1} \geqslant \left(\sum_{i=1}^{n} \left|dh^1 \wedge \dots \wedge dh^{i-1} \wedge dh^{i+1} \wedge \dots \wedge dh^n \wedge dt\right|^2\right)^{\frac{1}{2}} \tag{6.37}$$

Note that here in both sides only spherical derivatives of h are significant; the terms containing $h_N\, dt$ vanish after wedging them with dt. This observation permits us to replace dh^i by the covectors $\mathfrak{a}_i = dh^i - h^i_N dt$. We view $\mathfrak{a}_i$ as elements of the space $\bigwedge^1(\mathbb{R}^{n-1})$. Once this interpretation is accepted, the proof continues via an algebraic inequality.

LEMMA 6.2. *For every n-tuple of covectors in $\mathbb{R}^{n-1}$, $\mathfrak{a}_1, \dots, \mathfrak{a}_n \in \bigwedge^1(\mathbb{R}^{n-1})$, we have*

$$|\mathfrak{a}_1|^2 + \dots + |\mathfrak{a}_n|^2 \geqslant (n-1)\left(\sum_{i=1}^{n} |\mathfrak{a}_1 \wedge \dots \wedge \mathfrak{a}_{i-1} \wedge \mathfrak{a}_{i+1} \wedge \dots \wedge \mathfrak{a}_n|^2\right)^{\frac{1}{n-1}} \tag{6.38}$$

PROOF. We look at the matrix M of size $n \times (n-1)$ whose rows are made of the covectors $\mathfrak{a}_1, \dots, \mathfrak{a}_n$. Let $\mathfrak{b}_2, \dots, \mathfrak{b}_n \in \mathbb{R}^n$ denote the column vectors of M

$$\begin{bmatrix} - & - & \mathfrak{a}_1 & - & - \\ - & - & \mathfrak{a}_2 & - & - \\ & & \ddots & & \\ - & - & \mathfrak{a}_n & - & - \end{bmatrix} = M = \begin{bmatrix} | & | & & | \\ | & | & & | \\ \mathfrak{b}_2 & \mathfrak{b}_3 & \cdots & \mathfrak{b}_n \\ | & | & & | \\ | & | & & | \end{bmatrix} \tag{6.39}$$

In the left hand side of (6.38) we are actually dealing with the square of the Hilbert-Schmidt norm of M

$$|\mathfrak{a}_1|^2 + \dots + |\mathfrak{a}_n|^2 = \mathrm{Tr}\,(M^*M) = |\mathfrak{b}_2|^2 + \dots + |\mathfrak{b}_n|^2 \tag{6.40}$$

On the other hand, the sum in the right hand side of (6.38) is the Hilbert-Schmidt square of the norm of the cross product of the column vectors

$$\sum_{i=1}^{n} |\mathfrak{a}_1 \wedge \dots \wedge \mathfrak{a}_{i-1} \wedge \mathfrak{a}_{i+1} \wedge \dots \wedge \mathfrak{a}_n|^2 = |\mathfrak{b}_2 \times \dots \times \mathfrak{b}_n|^2 \leqslant |\mathfrak{b}_2|^2 \cdots |\mathfrak{b}_n|^2$$

by Hadamard's inequality. The inequality between arithmetic and geometric means yields

$$|\mathfrak{b}_2|^2 \cdots |\mathfrak{b}_n|^2 \leqslant \left(\frac{|\mathfrak{b}_2|^2 + \dots + |\mathfrak{b}_n|^2}{n-1}\right)^{n-1} \tag{6.41}$$

From here Lemma 6.2 is immediate. This also completes the proof of Proposition 6.1.

CHAPTER 7

Free Lagrangians

We are about to introduce the basic concept of this paper. Given two domains Ω and Ω^* in $\mathbb{R}^n$, we shall consider orientation preserving homeomorphisms $h : \Omega \to \Omega^*$ in a suitable Sobolev class $\mathscr{W}^{1,p}(\Omega, \Omega^*)$ so that a given energy integral

$$I[h] = \int_\Omega \mathcal{E}(x, h, Dh)\, dx \tag{7.1}$$

is well defined. The term *free Lagrangian* pertains to a differential n-form $\mathcal{E}(x, h, Dh)\, dx$ whose integral depends only on the homotopy class of $h : \Omega \xrightarrow{\text{onto}} \Omega^*$. The Jacobian determinant is pretty obvious example; we have,

$$\int_\Omega J(x, h)\, dx = |\Omega^*| \tag{7.2}$$

This identity holds for all orientation preserving homeomorphisms of Sobolev class $\mathscr{W}^{1,n}(\Omega, \Omega^*)$. Many more differential expressions enjoy a property such as this. In the next three lemmas we collect examples of free Lagrangians for homeomorphisms $h : \mathbb{A} \xrightarrow{\text{onto}} \mathbb{A}^*$ of annuli $\mathbb{A} = \{x\,;\quad r < |x| < R\}$ and $\mathbb{A}^* = \{x\,;\quad r_* < |x| < R_*\}$.

LEMMA 7.1. *Let $\Phi : [r_*, R_*] \to \mathbb{R}$ be any integrable function. Then the n-form*

$$\Phi(|h|)\, dh^1 \wedge ... \wedge dh^n \tag{7.3}$$

is a free Lagrangian. Precisely, we have

$$\int_{\mathbb{A}} \Phi(|h|) J(x, h)\, dx = \omega_{n-1} \int_{r_*}^{R_*} \tau^{n-1}\, \Phi(\tau)\, d\tau \tag{7.4}$$

for every orientation preserving homeomorphism $h \in \mathscr{W}^{1,n}(\mathbb{A}, \mathbb{A}^*)$.

This is none other than a general formula of integration by substitution.

LEMMA 7.2. *The following differential n-form*

$$\sum_{i=1}^{n} \frac{x_i\, dx_1 \wedge ... \wedge dx_{i-1} \wedge d|h| \wedge dx_{i+1} \wedge ... \wedge dx_n}{|h|\, |x|^n} = \frac{(d|h|) \wedge \star dt}{|h|\, t^{n-1}} \tag{7.5}$$

is a free Lagrangian in the class of all homeomorphisms $h \in \mathscr{W}^{1,1}(\mathbb{A}, \mathbb{A}^*)$ that preserve the order of the boundary components of the annuli $\mathbb{A}$ and $\mathbb{A}^*$. Precisely, we have

$$\int_{\mathbb{A}} \frac{d|h| \ \wedge \star dt}{|h|\, t^{n-1}} = \operatorname{Mod} \mathbb{A}^* \tag{7.6}$$

PROOF. First observe that the function $|h| : \mathbb{A} \to (r_*, R_*)$ extends continuously to the closure of $\mathbb{A}$ so that $|h(x)| = r_*$ for $|x| = r$ and $|h(x)| = R_*$ for $|x| = R$. The point is that our arguments will actually work for arbitrary continuous mappings $h : \overline{\mathbb{A}} \to \mathbb{R}^n_\circ$ of Sobolev class $\mathscr{W}^{1,1}(\mathbb{A}, \mathbb{R}^n)$ which satisfy the above boundary conditions, not necessarily homeomorphisms. The advantage in such a generality is that we can apply smooth approximation of h without worrying about injectivity. This extends Lemma 7.2 to all permissible mappings. If h is smooth it is legitimate to apply Stokes' theorem. Recalling the form ω and (6.26), we write the integral at (7.6) as:

$$\begin{aligned} \int_{\mathbb{A}} (d\log|h|) \wedge \omega = \int_{\mathbb{A}} d\,(\omega\, \log|h|) &= (\log R_*) \int_{|x|=R} \omega - (\log r_*) \int_{|x|=r} \omega \\ &= \omega_{n-1} \log \frac{R_*}{r_*} = \operatorname{Mod} \mathbb{A}^* \end{aligned} \tag{7.7}$$

The same proof works for more general integrals of the form

$$\int_{\mathbb{A}} \Phi'\left(|h|\right) \frac{d|h| \wedge \star dt}{t^{n-1}} = \omega_{n-1}\left[\Phi(R_*) - \Phi(r_*)\right] \tag{7.8}$$

with $\Phi \in \mathscr{C}^1[r_*, R_*]$. This also reads as

$$\int_{\mathbb{A}} \frac{\Phi'(\,|h|\,)\,|h|_N}{|x|^{n-1}} = \omega_{n-1}\left[\Phi(R_*) - \Phi(r_*)\right] \tag{7.9}$$

Another example of a free Lagrangian rests on the topological degree. Let $\Psi : S_t^{n-1} \to \mathbb{S}^{n-1}$ be any smooth mapping of a sphere $S_t^{n-1} = \{x\,;\ |x| = t\}$ into the unit sphere $\mathbb{S}^{n-1} \subset \mathbb{R}^n$. Then there is an integer, denoted by $\deg \Psi$, and called the *degree* of Ψ, such that

$$\int_{S_t^{n-1}} \Psi^\sharp(\omega) = \omega_{n-1} \deg \Psi \tag{7.10}$$

where ω stands for the standard surface measure of the unit sphere $\mathbb{S}^{n-1} \subset \mathbb{R}^n$, as given in (6.25). Note a general fact that $\deg \Psi$ remains unchanged under small perturbations of Ψ. If Ψ is an orientation-preserving diffeomorphism, then $\deg \Psi = 1$. Now a free Lagrangian somewhat dual to that in Lemma 7.2, is obtained as follows.

LEMMA 7.3. *The following differential n-form*

$$\sum_{i=1}^{n} \frac{h^i\, dh^1 \wedge \ldots \wedge dh^{i-1} \wedge d|x| \wedge dh^{i+1} \wedge \ldots \wedge dh^n}{|x|\,|h|^n} \tag{7.11}$$

is a free Lagrangian in the class of all orientation preserving homeomorphism mappings $h \in \mathscr{W}^{1,n-1}(\mathbb{A}, \mathbb{A}^*)$. Precisely, we have

$$\int_{\mathbb{A}} \frac{dt}{t} \wedge h^\sharp \omega = \operatorname{Mod} \mathbb{A} \tag{7.12}$$

PROOF. As a first step, we approximate h by smooth mappings h_ν converging to h c-uniformly on $\mathbb{A}$ and in the norm of $\mathscr{W}^{1,n-1}(\mathbb{A}, \mathbb{R}^n)$. For ν sufficiently large we consider the mappings

$$\Psi = \frac{h_\nu}{|h_\nu|} : \mathbb{A}' \to \mathbb{S}^{n-1}, \quad \nu = 1, 2, \ldots \tag{7.13}$$

defined on a slightly smaller annulus $\mathbb{A}' = \{x\,;\quad r' \leqslant |x| \leqslant R'\}$, where $r < r' < R' < R$. The degree of Ψ restricted to any concentric sphere in $\mathbb{A}'$ is equal to

1. Moreover, we have the point-wise identity $\Psi^\sharp\omega = h_\nu^\sharp(\omega)$. Now we integrate as follows

$$\int_{\mathbb{A}'} \frac{dt}{t} \wedge \Psi^\sharp\omega = \int_{\mathbb{A}'} d\left(\log|t|\, \Psi^\sharp\omega\right) \tag{7.14}$$

because $d\left(\Psi^\sharp\omega\right) = \Psi^\sharp\left(d\omega\right) = 0$. By Stokes' theorem, the right hand side is equal to:

$$\int_{|x|=R'} \log|x|\, \Psi^\sharp\omega - \int_{|x|=r'} \log|x|\, \Psi^\sharp\omega = \omega_{n-1}\log R' \,-\, \omega_{n-1}\log r' = \operatorname{Mod}\mathbb{A}'$$

Passing to the limit as $\nu \to \infty$ we conclude with the formula

$$\int_{\mathbb{A}'} \frac{dt}{t} \wedge h^\sharp\omega = \operatorname{Mod}\mathbb{A}' \tag{7.15}$$

Finally letting $r' \to r$ and $R' \to R$ yield (7.12).

REMARK 7.4. The reader may wish to generalize this lemma to mappings $h : \mathbb{A} \to \mathbb{R}^n_\circ$ which are not necessarily homeomorphisms. The degree of $\frac{h}{|h|}$ will emerge as a factor in front of the right hand side of (7.12).

The following corollary from Lemmas 7.1, 7.2, 7.3 provide us with three inequalities, which we call free Lagrangian estimates.

COROLLARY 7.5. *Let h be a homeomorphism between spherical rings $\mathbb{A}$ and $\mathbb{A}^*$ in the Sobolev class $\mathscr{W}^{1,n}(\mathbb{A},\mathbb{A}^*)$. Then*

$$\int_{\mathbb{A}} \Phi\big(|h|\big)\, |h_N|\, |h_T|^{n-1} \geqslant \omega_{n-1}\int_{r_*}^{R_*} \tau^{n-1}\Phi(\tau)\, d\tau \tag{7.16}$$

whenever Φ is integrable in $[r_*, R_*]$. We have the equality in (7.16) if and only if $|h_N|\, |h_T|^{n-1} = J(x,h)$. Furthermore,

$$\int_{\mathbb{A}} \frac{|h_N|}{|h|\, |x|^{n-1}} \geqslant \operatorname{Mod}\mathbb{A}^* \tag{7.17}$$

$$\int_{\mathbb{A}} \frac{|h_T|^{n-1}}{|h|^{n-1}\, |x|} \geqslant \operatorname{Mod}\mathbb{A} \tag{7.18}$$

Note that we have equalities if h is a radial mapping. Other cases of equalities are also possible.

CHAPTER 8

Some Estimates of Free Lagrangians

Before we proceed to the general proofs, it is instructive to take on some estimates which are at the heart of our application of free Lagrangians.

1. The $\mathscr{F}_h$-energy integral with operator norm

The extremal problem is remarkably simpler if we use the operator norm of the differential. Let h be a permissible map of $\mathbb{A}$ onto $\mathbb{A}^*$. The following point-wise inequality is straightforward,

$$|Dh| \geqslant \max\{|h_N|\,,\,|h_T|\} \tag{8.1}$$

Indeed, we have

$$|Dh| \geqslant \max\{|h_N|, |h_{T_2}|, ..., |h_{T_n}|\} \geqslant \max\{|h_N|\,,\,|h_T|\} \tag{8.2}$$

Certainly, equality occurs when h is a radial stretching. All that is needed is to apply the lower bounds at (7.17) and (7.18).

$$\mathscr{F}_h = \int_{\mathbb{A}} \frac{|Dh|^n}{|h|^n} \geqslant \max\left\{\int_{\mathbb{A}} \frac{|h_N|^n}{|h|^n}\,,\,\int_{\mathbb{A}} \frac{|h_T|^n}{|h|^n}\right\} \tag{8.3}$$

Here we estimate each term by Hölder's inequality,

$$\begin{aligned}\int_{\mathbb{A}} \frac{|h_N|^n}{|h|^n} &\geqslant \left(\int_{\mathbb{A}} \frac{|h_N|}{|h|\,|x|^{n-1}}\right)^n \left(\int_{\mathbb{A}} \frac{dx}{|x|^n}\right)^{1-n} \\ &\geqslant (\operatorname{Mod}\mathbb{A}^*)^n\,(\operatorname{Mod}\mathbb{A})^{n-1} = \alpha^n \operatorname{Mod}\mathbb{A}\end{aligned} \tag{8.4}$$

where

$$\alpha \stackrel{\text{def}}{=\!=} \frac{\operatorname{Mod}\mathbb{A}^*}{\operatorname{Mod}\mathbb{A}} \tag{8.5}$$

Similarly, we obtain

$$\begin{aligned}\int_{\mathbb{A}} \frac{|h_T|^n}{|h|^n} &\geqslant \left(\int_{\mathbb{A}} \frac{|h_T|^{n-1}}{|h|^{n-1}|x|}\right)^{\frac{n}{n-1}} \left(\int_{\mathbb{A}} \frac{dx}{|x|^n}\right)^{\frac{1}{1-n}} \\ &\geqslant (\operatorname{Mod}\mathbb{A})^{\frac{n}{1-n}}\,(\operatorname{Mod}\mathbb{A})^{\frac{1}{n-1}} = \operatorname{Mod}\mathbb{A}\end{aligned} \tag{8.6}$$

Substituting these two estimates into (8.3) we conclude with the following Theorem

THEOREM 8.1. *Let $\mathbb{A}$ and $\mathbb{A}^*$ be spherical rings in $\mathbb{R}^n$, $n \geqslant 2$. Then for every $h \in \mathscr{P}(\mathbb{A}, \mathbb{A}^*)$ we have*

$$\int_{\mathbb{A}} \frac{|Dh|^n}{|h|^n} \geqslant \max\{1, \alpha^n\} \operatorname{Mod}\mathbb{A} \tag{8.7}$$

As for the sharpness of this estimate we note that equality holds for the power stretching $h(x) = |x|^{\alpha-1}x$.

There are, however, other cases of equality in (8.7) if $\operatorname{Mod}\mathbb{A}^* \neq \operatorname{Mod}\mathbb{A}$.

2. Radial symmetry

Suppose that the radial stretching

$$h(x) = H(|x|)\,\frac{x}{|x|}\,, \qquad H \in \mathscr{AC}\,[r,R] \tag{8.8}$$

maps homeomorphically an annulus $\mathbb{A} = \{x;\ \ r < |x| < R\}$ onto $\mathbb{A}^* = \{y;\ \ r_* < |y| < R_*\}$, where $\operatorname{Mod}\mathbb{A}^* = \alpha \operatorname{Mod}\mathbb{A}$, with some $\alpha > 0$. We may assume that $r_* = r^\alpha$ and $R_* = R^\alpha$. Thus, in particular, h preserves the order of the boundary components. The power stretching $h^\alpha = |x|^{\alpha-1}x$ serves as an example of such a homeomorphism. We shall prove that h^α is in fact the minimizer of $\mathscr{F}_h$ among all radial stretchings.

PROPOSITION 8.2. *For each radial stretching we have*

$$\mathscr{F}_h = \int_{\mathbb{A}} \frac{\|\,Dh\,\|^n}{|h|^n} \geqslant \int_{\mathbb{A}} \frac{\|\,Dh^\alpha\,\|^n}{|h^\alpha|^n} = \left(\alpha^2+n-1\right)^{\frac{n}{2}} \operatorname{Mod}\mathbb{A} \tag{8.9}$$

Equality holds modulo rotation of $\mathbb{A}$.

PROOF. We express the energy of h as

$$\mathscr{F}_h = \int_{\mathbb{A}} \frac{\left(|h_N|^2 + (n-1)\,|h_T|^2\right)^{\frac{n}{2}}}{|h|^n} \tag{8.10}$$

where the radial and spherical derivatives of h are found in terms of the strain function H by using formulas (6.19);

$$|h_N| = \left|\dot{H}(t)\right| \qquad \text{and} \qquad |h_T| = \frac{|H(t)|}{t} \tag{8.11}$$

Integration in polar coordinates leads to a line integral

$$\begin{aligned} \mathscr{F}_h &= \omega_{n-1} \int_r^R \left(\frac{t^2\dot{H}^2}{H^2} + n - 1\right)^{\frac{n}{2}} \frac{dt}{t} \\ &= \omega_{n-1} \log\frac{R}{r} \fint \left(X^2+n-1\right)^{\frac{n}{2}}\, d\mu \end{aligned} \tag{8.12}$$

where the integral average is taken with respect to the measure $d\mu = \frac{dt}{t}$. The key is that $X(t) = \frac{t\dot{H}}{H}$ is a one dimensional free Lagrangian. Indeed, we see that the integral mean

$$\fint_r^R X\, d\mu = \frac{1}{\log\frac{R}{r}} \int_r^R \frac{\dot{H}(t)\,dt}{H(t)} = \frac{\log\frac{R_*}{r_*}}{\log\frac{R}{r}} = \alpha \tag{8.13}$$

does not depend on H. Next using Jensen's inequality for the convex function $X \to (X^2+n-1)^{\frac{n}{2}}$ yields

$$\begin{aligned} \mathscr{F}_h &\geqslant \left[\left(\fint_r^R X\, d\mu\right)^2 + n - 1\right]^{\frac{n}{2}} \operatorname{Mod}\mathbb{A} \\ &= \left[\alpha^2+n-1\right]^{\frac{n}{2}} \operatorname{Mod}\mathbb{A} \end{aligned} \tag{8.14}$$

as desired. Examining these arguments backwards we obtain the uniqueness statement.

As a corollary, letting α go to 0, we observe that among all radial deformations $h : \mathbb{A} \to \mathbb{S}^{n-1}$, the n-harminic energy

$$\mathscr{E}_h = \int_{\mathbb{A}} \| Dh \|^n \geqslant (n-1)^{\frac{n}{2}} \operatorname{Mod} \mathbb{A} \tag{8.15}$$

assumes its minimal value for the mapping $h(x) = \frac{x}{|x|}$. This will be later shown to be true in a larger class of mappings.

In dimensions $n = 2, 3$ the assertion of Proposition 8.2 remains true for all permissible mappings in the Sobolev class $\mathscr{W}^{1,n}(\mathbb{A}, \mathbb{A}^*)$.[1]

3. Proof of Theorem 1.14

The proof in case $n = 2$ is accomplished by an elegant use of Hölder's inequality,

$$\begin{aligned}
\mathscr{F}_h &= \int_{\mathbb{A}} \frac{\| Dh \|^2}{|h|^2} = \int_{\mathbb{A}} \left(\left| \frac{h_N}{h} \right|^2 + \left| \frac{h_T}{h} \right|^2 \right) \\
&\geqslant \left[\left(\int_{\mathbb{A}} \frac{|h_N|}{|x|\,|h|} \right)^2 + \left(\int_{\mathbb{A}} \frac{|h_T|}{|x|\,|h|} \right)^2 \right] \left(\int_{\mathbb{A}} \frac{dx}{|x|^2} \right)^{-1}
\end{aligned} \tag{8.16}$$

The lower bounds at (7.17) and (7.18) yield

$$\mathscr{F}_h \geqslant \left[(\operatorname{Mod} \mathbb{A}^*)^2 + (\operatorname{Mod} \mathbb{A})^2 \right] (\operatorname{Mod} \mathbb{A})^{-1} = \left(\alpha^2 + 1 \right) \operatorname{Mod} \mathbb{A} = \mathscr{F}_{h^\alpha}$$

as claimed.

The proof in dimension $n = 3$ involves application of Hölder's inequality twice

$$\begin{aligned}
\mathscr{F}_h &= \int_{\mathbb{A}} \frac{\| Dh \|^3}{|h|^3} = \int_{\mathbb{A}} \left(\left| \frac{h_N}{h} \right|^2 + 2 \left| \frac{h_T}{h} \right|^2 \right)^{\frac{3}{2}} \\
&\geqslant \left[\int_{\mathbb{A}} \left(\left| \frac{h_N}{h} \right|^2 + 2 \left| \frac{h_T}{h} \right|^2 \right) \frac{dx}{|x|} \right]^{\frac{3}{2}} \left(\int_{\mathbb{A}} \frac{dx}{|x|^3} \right)^{-\frac{1}{2}} \\
&\geqslant \left[\int_{\mathbb{A}} \left(\left| \frac{h_N}{h} \right| \frac{dx}{|x|^2} \right)^2 \left(\int_{\mathbb{A}} \frac{dx}{|x|^3} \right)^{-1} + 2 \int_{\mathbb{A}} \left| \frac{h_T}{h} \right|^2 \frac{dx}{|x|} \right]^{\frac{3}{2}} \left(\int_{\mathbb{A}} \frac{dx}{|x|^3} \right)^{-\frac{1}{2}}
\end{aligned}$$

Now using lower bounds at (7.17) and (7.18) we find that

$$\begin{aligned}
\mathscr{F}_h &\geqslant \left[(\operatorname{Mod} \mathbb{A}^*)^2 (\operatorname{Mod} \mathbb{A})^{-1} + 2 \operatorname{Mod} \mathbb{A} \right]^{\frac{3}{2}} (\operatorname{Mod} \mathbb{A})^{-\frac{1}{2}} \\
&= \left(\alpha^2 + 2 \right)^{\frac{3}{2}} \operatorname{Mod} \mathbb{A} = \mathscr{F}_{h^\alpha}
\end{aligned} \tag{8.17}$$

For the uniqueness we refer to Section 3 where such problems are dealt in all dimensions. The lower bounds of free Lagrangians, for the energy integrand $\frac{\| Dh \|^n}{|h|^n}$ in

[1] In the next chapter we shall extend this result to all dimensions, but it will require an upper bound of the modulus of $\mathbb{A}^*$.

higher dimensions, are more sophisticated. We found optimal bounds if $\operatorname{Mod}\mathbb{A}^* \leqslant \alpha_n \operatorname{Mod}\mathbb{A}$, where $1 < \alpha_n < \infty$ is a solution to the algebraic equation

$$\left(\alpha_n^2 + n - 1\right)^{n-2}\left(\alpha_n^2 - 1\right)^2 = \alpha_n^{2n} \;\Rightarrow\; \alpha_n < \sqrt{\frac{n-1}{n-3}}. \tag{8.18}$$

CHAPTER 9

Proof of Theorem 1.15

It suffices to consider homeomorphisms $h : \mathbb{A} \to \mathbb{A}^*$ of class $\mathscr{W}^{1,n}(\mathbb{A}, \mathbb{A}^*)$. Recall that the target annulus is not too fat. Precisely we assume that

$$\operatorname{Mod} \mathbb{A}^* < \alpha_n \operatorname{Mod} \mathbb{A} \tag{9.1}$$

where the critical factor $\alpha_n > 1$ is determined by the equation (8.18), which we write as

$$\frac{(\alpha_n^2 + n - 1)^{\frac{n-2}{2}} (\alpha_n^2 - 1)}{\alpha_n^n} = 1 \tag{9.2}$$

LEMMA 9.1. *Let $X, Y \geqslant 0$ and $1 \leqslant \alpha < \alpha_n$. Then*

$$a = a(\alpha) \stackrel{\text{def}}{=\!=} \frac{(\alpha^2 + n - 1)^{\frac{n-2}{2}} (\alpha^2 - 1)}{\alpha^n} < 1 \tag{9.3}$$

and, we have

$$\left[X^2 + (n-1)Y^2\right]^{\frac{n}{2}} \geqslant a\, X^n + b\, X Y^{n-1} \tag{9.4}$$

where

$$b = \frac{n\left(\alpha^2 + n - 1\right)^{\frac{n-2}{2}}}{\alpha} \tag{9.5}$$

Equality holds if and only if $X = \alpha Y$.

PROOF. Because of homogeneity we may assume that $Y = 1$; the case $Y = 0$ is obvious. We are reduced to proving an inequality with one variable $X \geqslant 0$,

$$\left[X^2 + (n-1)\right]^{\frac{n}{2}} \geqslant aX^n + bX \tag{9.6}$$

Consider the function

$$\varphi(X) = \left(X^2 + n - 1\right)^{\frac{n}{2}} - aX^n - bX \tag{9.7}$$

and its two derivatives

$$\varphi'(X) = n\left(X^2 + n - 1\right)^{\frac{n-2}{2}} X - naX^{n-1} - b \tag{9.8}$$

$$\varphi''(X) = n(n-1)\left[\left(X^2 + n - 1\right)^{\frac{n-4}{2}} (X^2 + 1) - aX^{n-2}\right] > 0 \tag{9.9}$$

This letter estimate is guaranteed by our hypothesis that $a < 1$. Thus φ is strictly convex and, therefore, has at most one critical point. The coefficient b has been defined above exactly in a way to ensure that $\varphi'(\alpha) = 0$. The coefficient a has been chosen so that $\varphi(\alpha) = 0$. This completes the proof of Lemma 9.1.

LEMMA 9.2. *Let $X, Y \geqslant 0$ and $0 \leqslant \alpha \leqslant 1$. Then*

$$[X^2 + (n-1)Y^2]^{\frac{n}{2}} \geqslant a\,Y^n + b\,XY^{n-1} \tag{9.10}$$

where

$$a = (n-1)\left(\alpha^2 + n - 1\right)^{\frac{n-2}{2}}\left(1 - \alpha^2\right) \tag{9.11}$$

and

$$b = n\alpha\left(\alpha^2 + n - 1\right)^{\frac{n-2}{2}} \tag{9.12}$$

Equality holds if and only if $X = \alpha Y$.

PROOF. Because of homogeneity we may assume that $Y = 1$, the case $Y = 0$ is obvious. We are reduced to the inequality with one variable $X \geqslant 0$,

$$[X^2 + (n-1)]^{\frac{n}{2}} \geqslant a + bX \tag{9.13}$$

Consider the function

$$\varphi(X) = \left(X^2 + n - 1\right)^{\frac{n}{2}} - bX \tag{9.14}$$

and its two derivatives

$$\varphi'(X) = n\left(X^2 + n - 1\right)^{\frac{n-2}{2}} X - b \tag{9.15}$$

$$\varphi''(X) = n(n-1)\left(X^2 + n - 1\right)^{\frac{n-4}{2}} (X^2 + 1) > 0 \tag{9.16}$$

As before, φ is strictly convex with one critical point. The coefficient b has been defined exactly to ensure that $\varphi'(\alpha) = 0$, whereas $a = \varphi(\alpha)$. This completes the proof of Lemma 9.2.

We now proceed to the proof of Theorem 1.15. Consider two cases.

1. The case of expanding pair

We apply Lemma 9.1 with $X = \frac{|h_N|}{|h|}$, $Y = \frac{|h_T|}{|h|}$ and $\alpha = \frac{\operatorname{Mod}\mathbb{A}^*}{\operatorname{Mod}\mathbb{A}} \geqslant 1$, to obtain the point-wise inequality

$$\begin{aligned}\frac{\|\,Dh\,\|^n}{|h|^n} &= \left(\frac{|h_N|^2}{|h|^2} + (n-1)\frac{|h_T|^2}{|h|^2}\right)^{\frac{n}{2}} \\ &\geqslant a\frac{|h_N|^n}{|h|^n} + b\frac{|h_N|\,|h_T|^{n-1}}{|h|^n}\end{aligned} \tag{9.17}$$

Then we integrate (9.17) over the annulus $\mathbb{A}$. For the last term we may apply the lower bound at (7.16). The first term in the right hand side of (9.17) needs some adjustments before using the lower bound at (7.17). These adjustments are easily accomplished by Hölder's inequality.

$$\begin{aligned}\int_{\mathbb{A}}\frac{\|\,Dh\,\|^n}{|h|^n} &\geqslant a\left(\int_{\mathbb{A}}\frac{|h_N|}{|h|\,|x|^{n-1}}\right)^n\left(\int_{\mathbb{A}}\frac{dx}{|x|^n}\right)^{1-n} + b\operatorname{Mod}\mathbb{A}^* \\ &\geqslant a\left(\operatorname{Mod}\mathbb{A}^*\right)^n\left(\operatorname{Mod}\mathbb{A}\right)^{1-n} + b\operatorname{Mod}\mathbb{A}^* \\ &= \left(a\alpha^n + b\alpha\right)\operatorname{Mod}\mathbb{A} \\ &= \left(\alpha^2 + n - 1\right)^{\frac{n}{2}}\operatorname{Mod}\mathbb{A}\end{aligned} \tag{9.18}$$

because we have equality at (9.6) for $X = \alpha$. We see that the right hand side is the energy of the power stretching

$$h^\alpha(x) = r^* r^{-\alpha} |x|^{\alpha-1} x\,, \qquad \alpha = \frac{\operatorname{Mod}\mathbb{A}^*}{\operatorname{Mod}\mathbb{A}} \tag{9.19}$$

2. The case of contracting pair

This time we use Lemma 9.2 with the same data X, Y, α as were used in the case of expanding pair; that is, with $X = \frac{|h_N|}{|h|}$, $Y = \frac{|h_T|}{|h|}$ and $\alpha = \frac{\operatorname{Mod}\mathbb{A}^*}{\operatorname{Mod}\mathbb{A}} \leqslant 1$. This gives the following point wise inequality

$$\begin{aligned} \frac{\| Dh \|^n}{|h|^n} &= \left(\frac{|h_N|^2}{|h|^2} + (n-1)\frac{|h_T|^2}{|h|^2} \right)^{\frac{n}{2}} \\ &\geqslant a \frac{|h_T|^n}{|h|^n} + b\, \frac{|h_N|\,|h_T|^{n-1}}{|h|^n} \end{aligned} \tag{9.20}$$

We integrate it over the annulus $\mathbb{A}$. For the last term we apply the lower bound at (7.16). As in previous case the first term in the right hand side of (9.20) needs some adjustments before using the lower bound at (7.18). These adjustments are easily accomplished by Hölder's inequality.

$$\begin{aligned} \int_{\mathbb{A}} \frac{\| Dh \|^n}{|h|^n} &\geqslant a \left(\int_{\mathbb{A}} \frac{|h_T|^{n-1}}{|x|\,|h|^{n-1}} \right)^{\frac{n}{n-1}} \left(\int_{\mathbb{A}} \frac{dx}{|x|^n} \right)^{\frac{-1}{n-1}} + b \operatorname{Mod}\mathbb{A}^* \\ &\geqslant a\, (\operatorname{Mod}\mathbb{A})^{\frac{n}{n-1}}\, (\operatorname{Mod}\mathbb{A})^{\frac{-1}{n-1}} + b \operatorname{Mod}\mathbb{A}^* \\ &= (a + b\,\alpha) \operatorname{Mod}\mathbb{A} \\ &= \left(\alpha^2 + n - 1\right)^{\frac{n}{2}} \operatorname{Mod}\mathbb{A} \end{aligned} \tag{9.21}$$

because we have equality in (9.13) for $X = \alpha$. We easily see that the right hand side is the energy of the power stretching

$$h^\alpha(x) = r^* r^{-\alpha} |x|^{\alpha-1} x\,, \qquad \alpha = \frac{\operatorname{Mod}\mathbb{A}^*}{\operatorname{Mod}\mathbb{A}} \tag{9.22}$$

3. Uniqueness

The reader might want to compare this proof with Chapter 13, in which the uniqueness question is dealt in greater generality. Let $h : \mathbb{A} \to \mathbb{A}^*$ be any extremal mapping. First observe that in both cases we have used a general inequality

$$J(x,h) \leqslant |h_N|\, |h_T|^{n-1} \tag{9.23}$$

For h to be extremal, we must have equality. In view of (6.11) the vectors $h_N, h_{T_2}, \ldots, h_{T_n}$ are mutually orthogonal and $|h_{T_2}| = \ldots = |h_{T_n}| = |h_T|$. Using the matrix representation of Dh at (6.5) we find the Cauchy-Green tensor of h to be a diagonal matrix

$$D^*h\, Dh = \begin{bmatrix} |h_N|^2 & 0 & \cdots & 0 \\ 0 & |h_T|^2 & \cdots & 0 \\ & & \ddots & \\ 0 & 0 & \cdots & |h_T|^2 \end{bmatrix} \tag{9.24}$$

Another necessary condition for h to be extremal is that $|h_N| = \alpha\,|h_T|$, because of the equality cases in Lemmas 9.1 and 9.2. In this way, we arrive at the Beltrami type system for the extremal mapping

$$D^*h\; Dh = J(x,h)^{\frac{2}{n}}\mathbf{K} \tag{9.25}$$

where $\mathbf{K}$ is a constant diagonal matrix

$$\mathbf{K} = \begin{bmatrix} \alpha^{1-\frac{1}{n}} & 0 & \cdots & 0 \\ 0 & \alpha^{-\frac{1}{n}} & \cdots & 0 \\ & & \ddots & \\ 0 & 0 & \cdots & \alpha^{-\frac{1}{n}} \end{bmatrix} \tag{9.26}$$

The power mapping $h^\alpha = h^\alpha(x)$ is one of the homeomorphic solutions to this system. It is well known and easy to verify that other solutions are obtained by composing this particular one with a conformal transformation [**7, 43, 44**]. Thus h takes the form

$$h = g \circ h^\alpha \tag{9.27}$$

where $g : \mathbb{A}^* \to \mathbb{A}^*$ is a conformal automorphism of the target annulus $\mathbb{A}^* = A(r_*, \mathbb{R}_*)$ onto itself. Thus, up to rotation $g(y) = y$ or $g(y) = r_* R_* \frac{y}{|y|^2}$. In either case h is a power stretching up to the rotation. The proof of uniqueness is complete.

A study of the extremal problems for the conformal energy

$$\mathscr{E}_h = \int_{\mathbb{A}} \| \, Dh(x) \, \|^n \, dx \tag{9.28}$$

is far more involved. Even when the solutions turn out to be radial stretchings, they no longer represent elementary functions such as $h^\alpha = |x|^{\alpha-1}x$. And this is not the only difficulty; there are new phenomena in case of non-injective solutions. In view of these concerns, it is remarkable that the method of free Lagrangians is still effective for non-injective solutions. First we examine planar mappings.

Concerning mappings minimizing the $\mathscr{L}^p$-norm of the gradient, we refer the interested reader to [**18**] and references there.

Part 2

The n-Harmonic Energy

CHAPTER 10

Harmonic Mappings between Planar Annuli, Proof of Theorem 1.8

Let $h : \mathbb{A} \to \mathbb{A}^*$ be a homeomorphism between annuli in the Sobolev class $\mathscr{W}^{1,2}(\mathbb{A}, \mathbb{A}^*)$. We view h as a complex valued function. Let us recall the formulas

$$\| Dh \|^2 = |h_N|^2 + |h_T|^2 \tag{10.1}$$

and

$$\det Dh = \operatorname{Im}\left(h_T \, \overline{h_N}\right) \leqslant |h_N| \, |h_T| \tag{10.2}$$

Case 1. **The expanding pair, $\frac{R}{r} \leqslant \frac{R^*}{r^*}$.**

Thus the target annulus is conformally fatter than the domain. We find a unique number $\omega \leqslant 0$ such that

$$\frac{R}{r} = \frac{R_* + \sqrt{R_*^2 - \omega}}{r_* + \sqrt{r_*^2 - \omega}} \tag{10.3}$$

Without loss of generality we may assume, by rescaling the annulus $\mathbb{A}$ if necessary, that

$$R = R_* + \sqrt{R_*^2 - \omega} \quad \text{and} \quad r = r_* + \sqrt{r_*^2 - \omega} \tag{10.4}$$

We begin with the inequality

$$\left(\frac{|h| \, |h_N|}{\sqrt{|h|^2 - \omega}} - |h_T| \right)^2 \geqslant 0 \tag{10.5}$$

Equivalently,

$$\begin{aligned}
\| Dh \|^2 &\geqslant \frac{-\omega}{|h|^2 - \omega} \, |h_N|^2 + \frac{2\,|h|}{\sqrt{|h|^2 - \omega}} \, |h_N| \, |h_T| \\
&\geqslant \frac{-\omega}{|h|^2 - \omega} \, (\, |h|_N \,)^2 + \frac{2\,|h|}{\sqrt{|h|^2 - \omega}} \, |h_N| \, |h_T| \\
&= -\omega \left\{ \left[\log\left(|h| + \sqrt{|h|^2 - \omega} \right) \right]_N \right\}^2 + \frac{2\,|h|}{\sqrt{|h|^2 - \omega}} \, |h_N| \, |h_T|
\end{aligned}$$

Here we have used an elementary fact that $|\, h_N \,| \geqslant |\, |h|_N \,|$, equality occurs if and only if $\frac{h_N}{h}$ is a real valued function. Let us integrate this estimate. We apply

Hölder's inequality to the first term and the estimate (7.16) to the second term,

$$\begin{aligned}
\int_{\mathbb{A}} \| Dh \|^2 &\geqslant -\omega \left[\int_{\mathbb{A}} \left[\log \left(|h| + \sqrt{|h|^2 - \omega} \right) \right]_N \frac{dx}{|x|} \right]^2 \cdot \left(\int_{\mathbb{A}} \frac{dx}{|x|^2} \right)^{-1} + \\
&\quad + 4\pi \int_{r_*}^{R_*} \frac{\tau^2}{\sqrt{\tau^2 - \omega}} \, d\tau \\
&= -\omega \left| 2\pi \log \frac{R_* + \sqrt{R_*^2 - \omega}}{r_* + \sqrt{r_*^2 - \omega}} \right|^2 \left(2\pi \log \frac{R}{r} \right)^{-1} + \\
&\quad + 2\pi \left[\tau \sqrt{\tau^2 - \omega} + \omega \log(\tau + \sqrt{\tau^2 - \omega}) \right]_{\tau = r_*}^{R_*} \\
&= 2\pi R_* \sqrt{R_*^2 - \omega} - 2\pi r_* \sqrt{r_*^2 - \omega} \qquad (10.6)
\end{aligned}$$

Elementary inspection reveals that equality holds for the Nitsche mapping

$$h(z) = \frac{1}{2} \left(z + \frac{\omega}{\bar{z}} \right)$$

Case 2. **The contracting pair,** $\frac{R_*}{r_*} < \frac{R}{r} \leqslant \frac{R_*}{r_*} + \sqrt{\frac{R_*^2}{r_*^2} - 1}$**.**

In this case the target annulus is conformally thinner than the domain, but not too thin. We express this condition by using the equation (10.3), where this time $0 < \omega \leqslant r_*^2$. Again we may assume that the relations at (10.4) hold. The same inequality (10.5) can be expressed in a somewhat different form

$$\| Dh \|^2 \geqslant \omega \left| \frac{h_T}{h} \right|^2 + 2 \, |h_N| \, |h_T| \sqrt{1 - \frac{\omega}{|h|^2}} \qquad (10.7)$$

As before, we apply Hölder's inequality, which together with the estimate (7.16) yields

$$\int_{\mathbb{A}} \| Dh \|^2 \geqslant \omega \left(\int_{\mathbb{A}} \frac{|h_T|}{|x| \, |h|} \right)^2 \cdot \left(\int_{\mathbb{A}} \frac{dx}{|x|^2} \right)^{-1} + 4\pi \int_{r_*}^{R_*} \sqrt{\tau^2 - \omega} \, d\tau$$

Next, the estimate (7.18) gives

$$\begin{aligned}
\int_{\mathbb{A}} \| Dh \|^2 &\geqslant \omega \left(2\pi \log \frac{R}{r} \right)^2 \cdot \left(2\pi \log \frac{R}{r} \right)^{-1} + 4\pi \int_{r_*}^{R_*} \sqrt{\tau - \omega} \, d\tau \\
&= 2\pi\omega \log \frac{R}{r} + 2\pi \left[\tau \sqrt{\tau^2 - \omega} - \omega \log(\tau + \sqrt{\tau^2 - \omega}) \right]_{\tau = r_*}^{R_*} \qquad (10.8)
\end{aligned}$$

In view of (10.4) we find that

$$\int_{\mathbb{A}} \| Dh \|^2 \geqslant 2\pi R_* \sqrt{R_*^2 - \omega} - 2\pi r_* \sqrt{r_*^2 - \omega} \qquad (10.9)$$

Again equality holds for the Nitsche mapping

$$h(z) = \frac{1}{2} \left(z + \frac{\omega}{\bar{z}} \right)$$

The borderline case. Taking $\omega = r_*^2$ we obtain what is called the critical Nitsche map with $R = R_* + \sqrt{R_*^2 - r_*^2}$ and $r = r_*$,

$$h^*(z) = \frac{1}{2} \left(z + \frac{r_*^2}{\bar{z}} \right) \qquad (10.10)$$

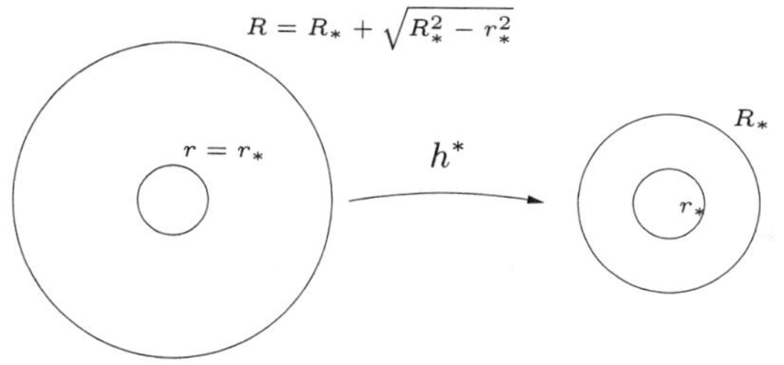

FIGURE 10.1. The Jacobian of h^* vanishes on the inner circle.

The conformal energy of the critical Nitsche map equals:

$$\int_{\mathbb{A}} \| Dh^* \|^2 = 2\pi R_* \sqrt{R_*^2 - r_*^2} \tag{10.11}$$

We are now in a position to consider the case:

Case 3. **Below the lower Nitsche bound,** $\frac{R_*}{r_*} + \sqrt{\frac{R_*^2}{r_*^2} - 1} < \frac{R}{r}$.

The target annulus $\mathbb{A}^*$ is too thin. We shall see that an inner part of $\mathbb{A}$ has to be hammered flat to the inner circle of $\mathbb{A}^*$. We can certainly rescale the annulus $\mathbb{A}$ to have $R = R_* + \sqrt{R_*^2 - r_*^2}$. This together with the hypothesis of this case, yields $r < r_*$.

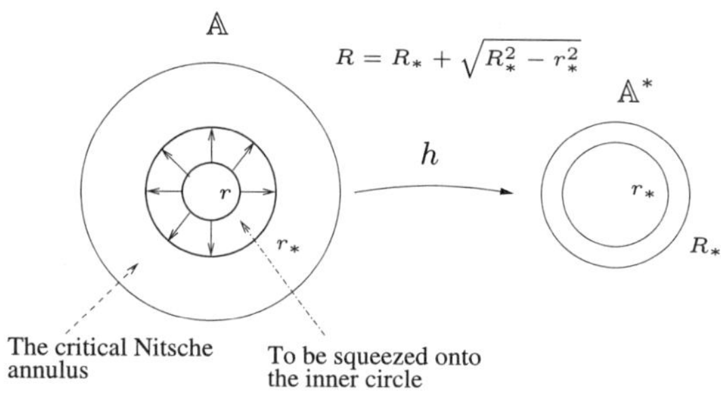

FIGURE 10.2. Below the lower Nitsche bound.

For every permissible map $h : \mathbb{A} \to \mathbb{A}^*$ we still have the estimate (10.8), in which we let $\omega = r_*^2$. Hence

$$\begin{aligned} \int_{\mathbb{A}} \| Dh \|^2 &\geqslant 2\pi r_* \log \frac{R}{r} + 2\pi \Big[\tau\sqrt{\tau^2 - r_*} - r_* \log(\tau + \sqrt{\tau^2 - r_*}) \Big]_{\tau = r_*}^{R_*} \\ &= 2\pi R_* \sqrt{R_*^2 - r_*^2} + 2\pi r_*^2 \log \frac{r_*}{r} \end{aligned} \tag{10.12}$$

The first term represents conformal energy of $h^* \colon \mathbb{A}(r_*, R) \to \mathbb{A}(r_*, R_*)$, by (10.11). The second term turns out to be exactly the energy of the hammering mapping $g(z) = r_* \frac{z}{|z|}$, which takes the remaining part $\mathbb{A}(r, r_*) \subset \mathbb{A}$ onto the inner circle of $\mathbb{A}(r_*, R_*)$.

$$\mathscr{E}_g = \int_{r<|z|<r_*} \| Dg \|^2 = 2\pi r_*^2 \log \frac{r_*}{r}. \tag{10.13}$$

Thus the energy of the mapping $h^\circ \colon \mathbb{A} \xrightarrow{\text{onto}} \mathbb{A}^*$

$$h^\circ = \begin{cases} r_* \frac{z}{|z|}, & r < |z| \leq r_* \\ \frac{1}{2}\left(z + \frac{r_*^2}{\bar{z}}\right), & r_* < |z| < R \end{cases}$$

is smaller than the energy of any homeomorphism $h \colon \mathbb{A}(r, R) \xrightarrow{\text{onto}} \mathbb{A}(r_*, R_*)$. It is clear that h° is a $\mathscr{W}^{1,2}$-limit of such homeomorphisms, completing the proof of Theorem 1.8.

Now we proceed to higher dimensions. Different estimates will be required for the contracting pairs of annuli than for the expanding pairs. Thus we devote a separate section for each case.

CHAPTER 11

Contracting Pair, $\operatorname{Mod}\mathbb{A}^* \leqslant \operatorname{Mod}\mathbb{A}$

First we consider the case when $\mathbb{A}^*$ is not too thin.

1. Proof of Theorem 1.12

Here we assume the Nitsche bound, $\mathcal{N}_{\dagger}(\operatorname{Mod}\mathbb{A}) \leqslant \operatorname{Mod}\mathbb{A}^*$, see (1.35) for $\mathcal{N}_{\dagger}$. This bound means precisely that there is a radial n-harmonic homeomorphism

$$h^\circ : \mathbb{A} \to \mathbb{A}^* , \qquad h^\circ(x) = H\big(|x|\big)\, \frac{x}{|x|} \tag{11.1}$$

Recall the characterictic equation at (5.28) for $H = H(t)$

$$\left[H^2 + \frac{t^2 \dot{H}^2}{n-1}\right]^{\frac{n-2}{2}} \Big(H^2 - t^2 \dot{H}^2\Big) \equiv c \tag{11.2}$$

where c is a positive constant determined uniquely by the rings $\mathbb{A}$ and $\mathbb{A}^*$. Equivalently,

$$\left(1 + \frac{\eta_{_H}^2}{n-1}\right)^{\frac{n-2}{2}} \big(1 - \eta_{_H}^2\big) = \frac{c}{H^n} \leqslant 1 \tag{11.3}$$

In particular, $c \leqslant [H(r)]^n = r_*^n$. Equation (11.3) suggests that we should consider the nonnegative solution $\eta = \eta(t)$ to the equation

$$\left(1 + \frac{\eta^2}{n-1}\right)^{\frac{n-2}{2}} (1 - \eta^2) = \frac{c}{t^n} \leqslant 1\,, \quad r_* < t < R_* \tag{11.4}$$

There is exactly one such solution. The values of η lie in the interval $[0, 1]$.

Now, let $h : \mathbb{A} \xrightarrow{\text{onto}} \mathbb{A}^*$ be any homeomorphism of annuli preserving both orientation and order of the boundary components. We assume that $h \in \mathscr{W}^{1,n}(\mathbb{A}, \mathbb{A}^*)$. For each $x \in \mathbb{A}$, we apply Lemma 9.2 with $X = |h_{_N}(x)|$, $Y = |h_{_T}(x)|$ and $\alpha = \eta\big(|h(x)|\big) \leqslant 1$, to obtain the point-wise inequality

$$\begin{aligned} \| \, Dh \, \|^n &= \Big[|h_N|^2 + (n-1)\, |h_T|^2\Big]^{\frac{n}{2}} \\ &\geqslant (n-1)^{\frac{n}{2}}\, c\, \frac{|h_T|^n}{|h|^n} + b\big(|h|\big)\, |h_N|\, |h_T|^{n-1} \end{aligned} \tag{11.5}$$

The coefficient $b\big(|h|\big)$ comes from (9.12) where we take $\alpha = \eta\big(|h|\big)$. An important fact about $b = b\big(|h|\big)$ is that we have equality at (11.5) if $|h_N| = \eta(|h|)\, |h_T|$. This is exactly happening for the radial n-harmonic map at (11.1), by the definition of the constant c. Let us integrate (11.5) over the ring $\mathbb{A}$. For the last term we may apply the lower bound at (7.16). For the first term in the right hand side of (11.5) we use Hölder's inequality, and then (7.18).

$$
\begin{aligned}
\int_{\mathbb{A}} \| Dh \|^n &\geqslant (n-1)^{\frac{n}{2}} c \left(\int_{\mathbb{A}} \frac{|h_T|^{n-1}}{|x|\,|h|^{n-1}} \right)^{\frac{n}{n-1}} \left(\int_{\mathbb{A}} \frac{dx}{|x|^n} \right)^{\frac{-1}{n-1}} + \\
&\quad + \omega_{n-1} \int_{r_*}^{R_*} \tau^{n-1} b(\tau)\, d\tau \\
&\geqslant (n-1)^{\frac{n}{2}} c \operatorname{Mod} \mathbb{A} + \omega_{n-1} \int_{r_*}^{R_*} \tau^{n-1} b(\tau)\, d\tau
\end{aligned}
\tag{11.6}
$$

Finally, observe that we have equalities in all estimates for the radial stretchings. Thus

$$
\int_{\mathbb{A}} \| Dh \|^n \geqslant \int_{\mathbb{A}} \| Dh^\circ \|^n \tag{11.7}
$$

as claimed. A proof of the uniqueness statement is postponed until Chapter 13. We only record further use that the equality at (11.7) yield

$$
|h_N|\, |h_T|^{n-1} = J(x,h) \tag{11.8}
$$

Next we turn to the case when $\mathbb{A}^*$ is significantly thinner than $\mathbb{A}$.

2. Proof of Theorem 1.13

Here we assume that the lower Nitsche bound fails; that is,

$$
\operatorname{Mod} \mathbb{A}^* < \mathcal{N}_\dagger(\operatorname{Mod} \mathbb{A})
$$

We split $\mathbb{A}$ into two concentric annuli

$$
\mathbb{A} = \mathbb{A}(r,R) = \mathbb{A}(r,1) \cup \mathbb{A}[1,R) \tag{11.9}
$$

where

$$
\operatorname{Mod} \mathbb{A}^* = \mathcal{N}_\dagger(\operatorname{Mod} \mathbb{A}[1,R)) \tag{11.10}
$$

Let $\aleph : \mathbb{A}[1,R) \to \mathbb{A}^*$ denote by critical Nitsche map. This is the radial n-harmonic function

$$
\aleph(x) = H(|x|) \frac{x}{|x|} \tag{11.11}
$$

determined by the characteristic equation

$$
\left[H^2 + \frac{t^2 \dot{H}^2}{n-1} \right]^{\frac{n-2}{2}} \left(H^2 - t^2 \dot{H}^2 \right) \equiv c \tag{11.12}
$$

Evaluating it at $t=1$ yields $c = [H(1)]^n = r_*^n$. Now consider any permissible map $h : \mathbb{A} \to \mathbb{A}^*$ in $\mathscr{P}(\mathbb{A}, \mathbb{A}^*)$. We may use inequality (11.6) with $c = r_*^n$ to obtain

$$
\int_{\mathbb{A}} \| Dh \|^n \geqslant (n-1)^{\frac{n}{2}} r_*^n \operatorname{Mod} \mathbb{A} + \omega_{n-1} \int_{r_*}^{R_*} \tau^{n-1} b(\tau)\, d\tau \tag{11.13}
$$

On the other hand we have the following identity for the critical Nitsche map

$$
\int_{\mathbb{A}[1,R]} \| D\aleph \|^n = (n-1)^{\frac{n}{2}} r_*^n \operatorname{Mod} \mathbb{A}[1,R) + \omega_{n-1} \int_{r_*}^{R_*} \tau^{n-1} b(\tau)\, d\tau \tag{11.14}
$$

Hence (11.13) takes the form

$$\int_{\mathbb{A}} \| Dh \|^n \geqslant (n-1)^{\frac{n}{2}} r_*^n \operatorname{Mod} \mathbb{A}(r,1) + \int_{\mathbb{A}[1,R)} \| D\aleph \|^n \tag{11.15}$$

As a final step we notice that the first term is precisely equal to the energy of the hammering map

$$(n-1)^{\frac{n}{2}} r_*^n \operatorname{Mod} \mathbb{A}(r,1) = \int_{\mathbb{A}(r,1)} \| Dg \|^n , \quad g(x) = r_* \frac{x}{|x|} \tag{11.16}$$

We now glue $\aleph$ and g along the sphere $|x| = 1$ to obtain a map $h^\circ : \mathbb{A} \to \mathbb{A}^*$

$$h^\circ = \begin{cases} g(x) & \text{on } \mathbb{A}(r,1) \\ \aleph(x) & \text{on } \mathbb{A}[1,R) \end{cases} \tag{11.17}$$

This map minimizes the conformal energy. Indeed,

$$\int_{\mathbb{A}} \| Dh \|^n \geqslant \int_{\mathbb{A}} \| Dh^\circ \|^n \tag{11.18}$$

for every $h \in \mathscr{P}(\mathbb{A}, \mathbb{A}^*)$. The question of uniqueness is discussed in Chapter 13.

CHAPTER 12

Expanding Pair, $\operatorname{Mod}\mathbb{A}^* > \operatorname{Mod}\mathbb{A}$

Such annuli determine uniquely a radial n-harmonic map $h^\circ : \mathbb{A} \to \mathbb{A}^*$ in the class $\mathcal{H}_-$,

$$h^\circ(x) = H\big(|x|\big)\,\frac{x}{|x|} \qquad \text{where } \mathcal{L}H \equiv c < 0 \tag{12.1}$$

If $\mathbb{A}^*$ is too fat, it will latter become clear that in dimensions $n \geqslant 4$ this radial n-harmonic mapping is not the minimum energy solution.

1. Within the bounds, $\operatorname{Mod}\mathbb{A} < \operatorname{Mod}\mathbb{A}^* \leqslant \mathcal{N}^\dagger(\operatorname{Mod}\mathbb{A})$

Here the upper Nitsche function $\mathcal{N}^\dagger = \mathcal{N}^\dagger(t)$, $t > 0$, is determined uniquely by requiring that the inequality $\operatorname{Mod}\mathbb{A}^* \leqslant \mathcal{N}^\dagger(\operatorname{Mod}\mathbb{A})$ be equivalent to

$$\frac{\big(\eta_{_H}^2 + n - 1\big)^{\frac{n-2}{2}}\big(\eta_{_H}^2 - 1\big)}{\eta_{_H}^n} \leqslant 1 \tag{12.2}$$

where $\eta_{_H} = \frac{t\dot H}{H}$ is the elasticity function of $h^\circ : \mathbb{A} \to \mathbb{A}^*$. A somewhat explicit formula for $\mathcal{N}^\dagger$ is given in (12.13) after an analysis of (12.2). It may be worth noting in advance that $\mathcal{N}^\dagger \equiv \infty$ in dimensions $n = 2, 3$. Thus the condition $\operatorname{Mod}\mathbb{A}^* \leqslant \mathcal{N}^\dagger(\operatorname{Mod}\mathbb{A})$ is actually void for $n = 2, 3$. By the definition of α_n at (9.2), condition (12.2) is equivalent to

$$\eta_{_H}(t) \leqslant \alpha_n\,, \qquad \text{for all } r \leqslant t \leqslant R \tag{12.3}$$

Since we are in the expanding case the function $\eta_{_H}$ is decreasing. We therefore need only assume that

$$1 < \eta_{_H}(r) \leqslant \alpha_n \tag{12.4}$$

Recall that $\alpha_2 = \alpha_3 = \infty$, so this condition poses no restriction on $\mathbb{A}^*$ in dimensions $n = 2$ and 3.

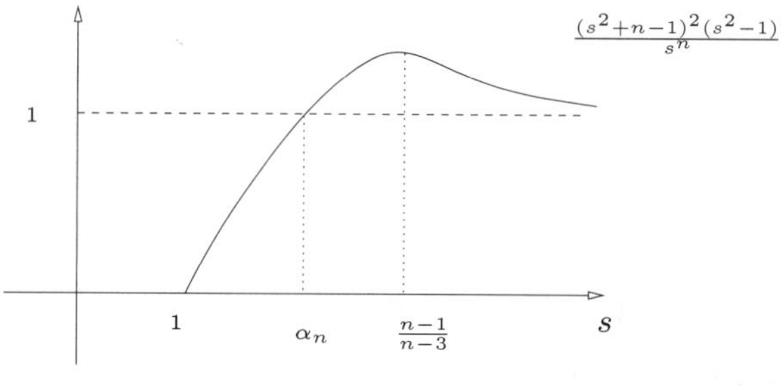

FIGURE 12.1. The bounds for the elasticity of h°.

Although it is not immediately clear the condition (12.4) is in fact a condition on the moduli of $\mathbb{A}$ and $\mathbb{A}^*$ alone, precisely

$$H_-(\gamma_n) \cdot \tfrac{R_*}{r_*} \leqslant H_-\left(\gamma_n \tfrac{R}{r}\right) \tag{12.5}$$ [1]

where the number $\gamma_n > 1$ (for $n \geqslant 4$) is determined by the equation

$$\gamma_n = \Gamma_-\left(\tfrac{1}{\alpha_n}\right) \tag{12.6}$$

Proof of (12.5). Since h° lies in the class $\mathcal{H}_-$ its strain function takes the form

$$H(t) = \lambda\, H_-(kt)\,, \qquad k > \frac{1}{r}\,, \qquad \text{for } r \leqslant t \leqslant R \tag{12.7}$$

The boundary constrains $H(r) = r_*$ and $H(R) = R_*$ yield a system of equations for the parameters λ and k

$$\lambda\, H_-(kr) = r_* \qquad \text{and} \qquad \lambda\, H_-(kR) = R_* \tag{12.8}$$

Once we eliminate λ the parameter k is determined (uniquely) from the equation

$$\frac{R_*}{r_*} = \frac{H_-(kR)}{H_-(kr)} \tag{12.9}$$

To solve this equation we look at the function

$$Q(k) = \frac{H_-(kR)}{H_-(kr)}\,, \qquad k > \frac{1}{r} \tag{12.10}$$

see (5.64). Elementary computation shows that

$$\frac{k\,\dot{Q}(k)}{Q(k)} = \eta_{\_}(Rk) - \eta_{\_}(rk) < 0 \tag{12.11}$$

In view of (5.65) we see that $Q(k)$ decreases from $+\infty$ to $\frac{R}{r}$ as k runs from $\frac{1}{r}$ to ∞. Recall that we are in the expanding case, $\frac{R_*}{r_*} > \frac{R}{r}$. Then by Mean Value Theorem there is unique $k > \frac{1}{r}$ satisfying (12.9). We now rewrite condition (12.6) as follows:

$$\begin{aligned} \eta_{_H}(r) &\leqslant \alpha_n \quad \Leftrightarrow \quad \eta_{\_}(k\,r) \leqslant \alpha_n \quad \Leftrightarrow \\ k\,r &\geqslant \eta_{\_}^{-1}(\alpha_n) = \Gamma_-\left(\tfrac{1}{\alpha_n}\right) = \gamma_n \quad \Leftrightarrow \quad k \geqslant \frac{\gamma_n}{r} \end{aligned} \tag{12.12}$$

Here we have used the identity $\Gamma_-(s) = \gamma_{\_}^{-1}\left(\frac{1}{s}\right)$, see (5.49) and (5.53). The latter inequality is equivalent to $Q\left(\frac{\gamma_n}{r}\right) \geqslant Q(k) = \frac{R_*}{r_*}$, which is the same as (12.5). We are now in a position to define

$$\mathcal{N}^\dagger(t) = \omega_{n-1} \log H_-\left(\gamma_n \exp\frac{t}{\omega_{n-1}}\right) - \omega_{n-1} \log H_-(\gamma_n). \tag{12.13}$$

PROPOSITION 12.1. *Let $h : \mathbb{A} \to \mathbb{A}^*$ be a permissible map in $\mathscr{P}(\mathbb{A}, \mathbb{A}^*)$ where*

$$\frac{R}{r} < \frac{R_*}{r_*} < \frac{H_-\left(\gamma_n \frac{R}{r}\right)}{H_-(\gamma_n)} \tag{12.14}$$

Then

$$\int_{\mathbb{A}} \| Dh \|^n \geqslant \int_{\mathbb{A}} \| Dh^\circ \|^n \tag{12.15}$$

Equality holds if and only if $h(x) = h^\circ(x)$ up to a conformal automorphism of $\mathbb{A}$.

---

[1] In dimensions $n = 2, 3$ we have $\alpha_n = \infty$, hence $\gamma_n = 1$. In this case $H_-(\gamma_n) = 0$ and, therefore, we impose no upper bound for Mod $\mathbb{A}^*$.

PROOF. The characteristic equation (5.28) for the map $h^\circ : \mathbb{A} \to \mathbb{A}^*$ defines a positive constant $q = -(n-1)^{\frac{n-2}{2}} \mathcal{L}H$; that is,

$$q \equiv \left[t^2\dot{H}^2 + (n-1)H^2\right]^{\frac{n-2}{2}} \left(t^2\dot{H}^2 - H^2\right) > 0 \tag{12.16}$$

Equivalently,

$$\left(n-1+\eta_H^2\right)^{\frac{n-2}{2}} \left(\eta_H^2 - 1\right) = \frac{q}{H^n} \tag{12.17}$$

Now, consider an arbitrary permissible map $h : \mathbb{A} \to \mathbb{A}^*$. We may assume that h is a homeomorphism preserving orientation and the order of the boundary components of the annuli. We introduce a function $\eta = \eta(t)$ implicitly defined for $r_* \leqslant t \leqslant R_*$ by the equation,

$$\left(n-1+\eta^2\right)^{\frac{n-2}{2}} \left(\eta^2 - 1\right) = \frac{q}{t^n} \tag{12.18}$$

Note that the equation (12.17) for η_H at $t = r$ coincides with (12.18) for η at $t = r_*$. Hence $\eta(r_*) = \eta_H(r)$. Since $\eta = \eta(t)$ is decreasing we find that for every $r_* \leqslant t \leqslant R_*$,

$$1 < \eta(t) \leqslant \eta(r_*) = \eta_H(r) \leqslant \alpha_n \tag{12.19}$$

Next we recall the function $a = a(\alpha)$, $1 \leqslant \alpha < \infty$, from the formula (9.3). By the definition of α_n at (8.18) it follows that

$$a(\eta) = \frac{\left(n-1+\eta^2\right)^{\frac{n-2}{2}} \left(\eta^2-1\right)}{\eta^n} \leqslant 1 \qquad \text{for all } r_* < t < R_* \tag{12.20}$$

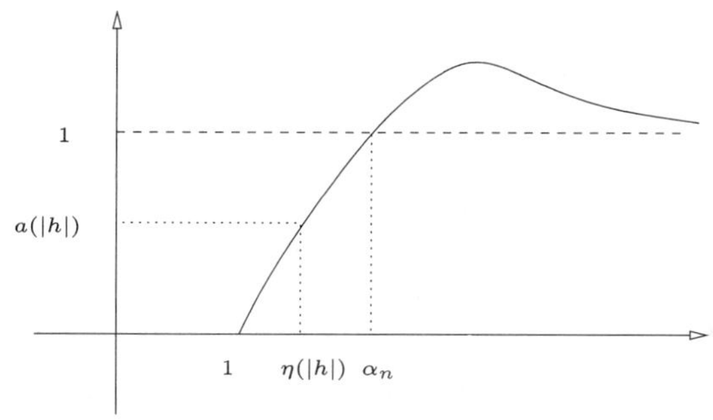

FIGURE 12.2. The coefficient $a(|h|)$.

Now Lemma 9.1 applies to $X = |h_N|$, $Y = |h_T|$ and $\alpha = \eta(|h|)$,

$$\begin{aligned} \| Dh \|^n &= \left[|h_N|^2 + (n-1)\,|h_T|^2\right]^{\frac{n}{2}} \\ &\geqslant a(|\eta|)\,|h_N|^n + b(|\eta|)\,|h_N|\,|h_T|^{n-1} \\ &= q\left[\frac{|h_N|}{|h|\,\eta(|h|)}\right]^n + \Phi(|h|)\,|h_N|\,|h_T|^{n-1} \end{aligned} \tag{12.21}$$

where $\Phi(|h|) = b\big(\eta(|h|)\big)$ and b is given by (9.5). According to Lemma 9.1, equality holds at a given point x if and only if $|h_N(x)| = \eta\big(|h(x)|\big)\,|h_T(x)|$. In particular, it holds almost everywhere for $h = h^\circ(x)$, because $|h^\circ_N| = \eta_H\, |h^\circ_T|$. We now integrate over the annulus $\mathbb{A}$. The last term at (12.21) is easily handled by the lower bound at (7.16),

$$\begin{aligned}\int_{\mathbb{A}} \Phi(|h|)\, |h_N|\, |h_T|^{n-1} &\geqslant \omega_{n-1} \int_{r_*}^{R_*} \tau^{n-1}\Phi(\tau)\, d\tau \\ &= \int_{\mathbb{A}} \Phi(|h^\circ|)\, |h^\circ_N|\, |h^\circ_T|^{n-1} \end{aligned} \tag{12.22}$$

The first term in the right hand side of (12.21) will be treated by Hölder's inequality in order to apply the lower bound at (7.17). This should be done in a way so one obtains equality for h°. Taking into an account the identities

$$\frac{|h^\circ_N|}{|h^\circ|\; \eta(|h^\circ|)} = \frac{\dot{H}}{H\, \eta_{_H}} = \frac{1}{|x|} \tag{12.23}$$

We proceed as follows

$$\int_{\mathbb{A}} \left[\frac{|h_N|}{|h|\; \eta(|h|)}\right]^n \geqslant \left(\int_{\mathbb{A}} \frac{|h|_N\; dx}{|h|\; \eta(|h|)\; |x|^{n-1}}\right)^n \left(\int_{\mathbb{A}} \frac{dx}{|x|^n}\right)^{1-n} \tag{12.24}$$

The first term in the right hand side is none other than a free Lagrangian at (7.9). This allows us to quickly evaluate the first term, without getting buried under massive computation.

$$\int_{\mathbb{A}} \frac{|h|_N\; dx}{|h|\; \eta(|h|)\; |x|^{n-1}} = \int_{\mathbb{A}} \frac{|h^\circ|_N\; dx}{|h^\circ|\; \eta(|h^\circ|)\; |x|^{n-1}} = \int_{\mathbb{A}} \frac{\dot{H}(|x|)\; dx}{H\, \eta_{_H}\; |x|^{n-1}} = \int_{\mathbb{A}} \frac{dx}{|x|^n}$$

Hence

$$\int_{\mathbb{A}} \left[\frac{|h_N|}{|h|\; \eta(|h|)}\right]^n \geqslant \int_{\mathbb{A}} \frac{dx}{|x|^n} = \operatorname{Mod} \mathbb{A} \tag{12.25}$$

In conclusion,

$$\int_{\mathbb{A}} \|\, Dh \,\|^{\,n} \geqslant q \operatorname{Mod} \mathbb{A} + \omega_{n-1} \int_{r_*}^{R_*} \tau^{n-1} \Phi(\tau)\, d\tau \tag{12.26}$$

with equality occuring for h°, as claimed.

The uniqueness results will be proven in more unified manner in Chapter 13.

1.1. Proof of Theorem 1.10. Indeed, the condition $\operatorname{Mod} \mathbb{A}^* \leqslant \mathcal{N}^\dagger(\operatorname{Mod} \mathbb{A})$ is equivalent to (12.14), in view of the explicit formula for $\mathcal{N}^\dagger$ given in (12.13).

1.2. Proof of Theorem 1.9. The proof is immediate from Proposition 12.1 once we recall that $\mathcal{N}^\dagger \equiv \infty$, for $n = 2, 3$ posing no upper bound for $\operatorname{Mod} \mathbb{A}^*$.

CHAPTER 13

The Uniqueness

There is a comprehensive approach to all our questions concerning uniqueness of the conformal energy

$$\mathscr{E}_h = \int_{\mathbb{A}} \| Dh(x) \|^n \mathrm{d}x \quad \text{-minimal mappings } h : \mathbb{A} \to \mathbb{A}^*. \tag{13.1}$$

We have already found that within the Nitsche bounds

$$\mathscr{N}_{\dagger}(\operatorname{Mod}\mathbb{A}) \leqslant \operatorname{Mod}\mathbb{A}^* \leqslant \mathscr{N}^{\dagger}(\operatorname{Mod}\mathbb{A}) \tag{13.2}$$

the radial n-harmonics

$$h^\circ(x) = H\big(|x|\big)\,\frac{x}{|x|} \tag{13.3}$$

are among the extremal solutions. We shall show that

THEOREM 13.1. *Below the upper Nitsche bound; that is,*

$$\operatorname{Mod}\mathbb{A}^* \leqslant \mathscr{N}^{\dagger}(\operatorname{Mod}\mathbb{A}) \tag{13.4}$$

every permissible minimizer $h : \mathbb{A} \to \mathbb{A}^*$ coincides with the radial extremal map modulo conformal automorphisms of $\mathbb{A}$.

Let $h \in \mathscr{P}(\mathbb{A}, \mathbb{A}^*)$ be any permissible extremal mapping. In all the preceding cases we came to the following equation as one of the necessary conditions for h to minimize the energy

$$|h_N| = \eta\big(|h|\big)\ |h_T| \tag{13.5}$$

see (9.17), (9.20), (11.5) and (12.21). Here $\eta = \eta(\tau)$ is a nonnegative function defined for $r_* \leqslant \tau \leqslant R_*$ by the rule

$$\tau^n \left(\eta^2 + n - 1\right)^{\frac{n-2}{2}} \left(\eta^2 - 1\right) \equiv c \tag{13.6}$$

Here the constant c comes from the characteristic equation for the radial extremal map $h^\circ(x) = H\big(|x|\big)\,\frac{x}{|x|}$. Precisely, we have

$$c \equiv [H(t)]^n \left(\eta_{_H}^2 + n - 1\right)^{\frac{n-2}{2}} \left(\eta_{_H}^2 - 1\right), \qquad \text{where } \eta_{_H}(t) = \frac{t\,\dot{H}(t)}{H(t)} \tag{13.7}$$

Note that η is strictly positive if $c > 0$. However, if $c < 0$ then $\eta(\tau)$ is strictly increasing and, therefore, can vanish only at the endpoint $\tau = r_*$. This latter situation arises when $\eta_{_H} \equiv 0$ in the hammering part of h°. Another necessary condition for a mapping to be extremal takes the form

$$|h_N|\ |h_T|^{n-1} = J(x,h) \tag{13.8}$$

because in each case we used the estimate (7.16) , which is sharp only when (13.8) holds. Consequently, the Cauchy-Green tensor of h becomes a diagonal matrix,

$$D^*h \cdot Dh = \begin{bmatrix} |h_N|^2 & 0 & \cdots & 0 \\ 0 & |h_T|^2 & \cdots & 0 \\ & & \ddots & \\ 0 & 0 & \cdots & |h_T|^2 \end{bmatrix} \stackrel{\text{def}}{=} \mathbf{C}(x) \tag{13.9}$$

see (6.12). Whenever Hölder's inequalities were used we always arrived at one of the following conditions for h to become extremal

$$|\, h_T \,| = \frac{C_1\,|h|}{|x|}\,, \quad \text{or} \quad |\, h_N \,| = \frac{C_2\,|h|}{|x|}\eta\big(|h|\big) \tag{13.10}$$

where C_1 and C_2 are constants, see (11.6) and (12.24). No matter which case we look at, the conclusion is that

$$|\, h_T \,| = \frac{\lambda\,|h|}{|x|} \quad \text{and} \quad |\, h_N \,| = \frac{\lambda\,|h|}{|x|}\eta\big(|h|\big) \tag{13.11}$$

for some constant $\lambda > 0$, because of (13.5). This constant λ is the same for all extremal solutions; in fact, $\lambda = 1$. This can be observed as follows. Since equality holds in either (11.6) or (12.25), at least one of the integrals

$$\int_{\mathbb{A}} \frac{|h_T|^{n-1}}{|x|\,|h|^{n-1}} \quad \text{and} \quad \int_{\mathbb{A}} \left[\frac{|h_N|}{|x|\,\eta(|h|)}\right]^n$$

is equal to $\operatorname{Mod}\mathbb{A}$. In view of (13.11) this yields $\lambda = 1$. Hence

$$D^*h \cdot Dh = \frac{\lambda^2\,|h|^2}{|x|^2} \begin{bmatrix} \eta^2\big(|h|\big) & 0 & \cdots & 0 \\ 0 & 1 & \cdots & 0 \\ & & \ddots & \\ 0 & 0 & \cdots & 1 \end{bmatrix} = \mathbf{G}(x, |h|) \tag{13.12}$$

We emphasize that the metric tensor $\mathbf{G} = \mathbf{G}(x, \tau)$, viewed as a function of two variables $x \in \mathbb{A}$ and $\tau \in (r_*, R_*)$, is the same for all extremals. A key step in establishing uniqueness is:

1. The point Cauchy problem

We shall consider the Cauchy-Green equation

$$D^*h(x)\, Dh(x) = \mathbf{G}(x, |h|) \tag{13.13}$$

for mappings $h : \Omega \to \mathbb{A}^*$ defined in a domain $\Omega \subset \mathbb{R}^n$ and valued in an annulus $\mathbb{A}^* = \mathbb{A}\,(r_*, R_*) \subset \mathbb{R}^n$. Here the function

$$\mathbf{G} : \Omega \times (r_*, R_*) \to \mathbb{R}^{n\times n}_+ = \begin{cases} \text{the space of symmetric} \\ \text{positive definite matrices} \end{cases} \tag{13.14}$$

is assumed to be $\mathscr{C}^\infty$-smooth. As for the regularity of the solutions, we initially assume that h is only continuous. It then follows from the formula

$$\|\, Dh \,\|^2 = \operatorname{Tr} \mathbf{G}(x, |h|) \in \mathscr{L}^\infty_{\text{loc}}(\Omega) \tag{13.15}$$

that h is actually locally Lipschitz continuous. One can look at Ω as a Riemannian manifold equipped with the positive definite element of arclength

$$\mathrm{d}s^2 = G_{ij}(x)\, \mathrm{d}x^i \otimes \mathrm{d}x^j \qquad \text{where} \quad [G_{ij}(x)] = \mathbf{G}(x, |h(x)|)$$

In this way $h:\Omega\to\mathbb{A}^*$ becomes a local isometry with respect to this metric tensor on Ω. At this point we may appeal to the well known regularity result due to Calabi and Hartman [**8**]. It tells us that if $D^*h\cdot Dh\in\mathscr{C}^{k,\alpha}(\Omega,\mathbb{R}^{n\times n}_+)$ for some $0<\alpha<1$ and $k=0,1,...$, then $h\in\mathscr{C}^{k+1,\alpha}(\Omega,\mathbb{R}^n)$. This result can be applied repeatedly to infer that in fact $h\in\mathscr{C}^\infty(\Omega,\mathbb{A}^*)$. One more observation is that the system (13.13) is invariant under linear isometries of the target annulus $\mathbb{A}^*$. Precisely, if h solves this system so does the mapping Th for every linear isometry $T:\mathbb{R}^n\to\mathbb{R}^n$. A priori the system (13.13) may not admit any local solution. But if it does, the Riemann curvature tensor of $\mathbf{G}(x,|h(x)|)$ must vanish. We shall take advantage of the classical computation of curvature. We express the second derivatives of h in terms of its first order derivatives

$$\frac{\partial h}{\partial x_j\,\partial x_k}=\sum_{\nu=1}^{n}\Gamma^\nu_{jk}\frac{\partial h}{\partial x_\nu} \tag{13.16}$$

Here Γ^ν_{jk} are the Christoffel symbols; explicitly,

$$\Gamma^\nu_{jk}=\frac{1}{2}\sum_{i=1}^{n}G^{i\nu}\left(\frac{\partial G_{ij}}{\partial x_k}+\frac{\partial G_{ik}}{\partial x_j}-\frac{\partial G_{jk}}{\partial x_i}\right) \tag{13.17}$$

where $G^{i\nu}$ are the entries of the inverse matrix to $\mathbf{G}$, see for instance [**29**, p. 37]. It should be noted that the partials $\frac{\partial\mathbf{G}}{\partial x_k}$ are computed in accordance with the chain rule; so in fact we have

$$\frac{\partial\mathbf{G}}{\partial x_k}=\mathbf{G}_{x_k}+\mathbf{G}_\tau\frac{\partial|h|}{\partial x_k} \tag{13.18}$$

where $\mathbf{G}_{x_k}$ and $\mathbf{G}_\tau$ stand for the partial derivatives of the function $(x,\tau)\to\mathbf{G}(x,\tau)$. Thus, in particular, the Christoffel symbols depend linearly on Dh. Let us state the general form of the second order equations obtained in this way

$$D^2h(x)=\Phi(x,\,|h|,\,Dh) \tag{13.19}$$

where $\Phi:\Omega\times(r_*,R_*)\times\mathbb{R}^{n\times n}\to\mathbb{R}^{n\times n\times n}$ is a given $\mathscr{C}^\infty$-smooth function. This is in fact a quadratic polynomial with respect to Dh. It should be noted that (13.19) is frame indifferent; that is, it holds for Th if it holds for h. With these equations at hand we can now prove the following

LEMMA 13.2. (UNIQUENESS IN THE POINT CAUCHY PROBLEM) *Suppose we are given two solutions h° and h to the Cauchy-Green equation (*13.13*) such that $|h(a)|=|h^\circ(a)|$ for some point $a\in\Omega$. Then there is an isometry $T:\mathbb{R}^n\to\mathbb{R}^n$ such that*

$$h(x)=T\,h^\circ(x)\qquad \textit{for all }\ x\in\Omega \tag{13.20}$$

PROOF. It follows from the equation (13.13) that

$$D^*h(a)\,Dh(a)=D^*h^\circ(a)\,Dh^\circ(a) \tag{13.21}$$

Therefore, there is an isometry $T:\mathbb{R}^n\to\mathbb{R}^n$ such that

$$Dh(a)=T\circ Dh^\circ(a)=D\big(Th^\circ\big)(a)$$

With a suitable choice of the isometry we may assume without loss of generality that the first order derivatives of h and h° also coincide at a; namely,

$$|h(a)|=|h^\circ(a)|\qquad\text{and}\qquad Dh(a)=Dh^\circ(a) \tag{13.22}$$

We are going to show that these two equations hold in the entire domain Ω. Obviously, the set where (13.22) hold is relatively closed in Ω. Thus, we need only show that this set is also open. To this end we consider a small ball

$$\mathbb{B} = \{\, x \in \mathbb{R}^n \,; \quad |x-a| \leqslant \epsilon \,\}$$

center at a and contained in Ω. We estimate the supremum norms of $Dh - Dh^\circ$ over $\mathbb{B}$ with ϵ approaching zero.

$$\begin{aligned} \| Dh - Dh^\circ \|_{\mathscr{L}^\infty(\mathbb{B})} &\preccurlyeq \epsilon \, \| D^2h - D^2h^\circ \|_{\mathscr{L}^\infty(\mathbb{B})} \\ &= \epsilon \, \| \Phi(x, |h|, Dh) - \Phi(x, |h^\circ|, Dh^\circ) \|_{\mathscr{L}^\infty(\mathbb{B})} \\ &\preccurlyeq \epsilon \, \| \, |h| - |h^\circ| \, \|_{\mathscr{L}^\infty(\mathbb{B})} + \epsilon \, \| \, Dh - Dh^\circ \, \|_{\mathscr{L}^\infty(\mathbb{B})} \end{aligned}$$

Here and subsequently, the symbol $\preccurlyeq$ stands for the inequality with a constant independent of ϵ. This constant varies from line to line. Noting that

$$\epsilon \, \| \, |h| - |h^\circ| \, \|_{\mathscr{L}^\infty(\mathbb{B})} \preccurlyeq \epsilon \, \| \, Dh - Dh^\circ \, \|_{\mathscr{L}^\infty(\mathbb{B})} \,, \tag{13.23}$$

we conclude that $Dh = Dh^\circ$ and $|h| = |h^\circ|$ in $\mathbb{B}$, as desired.

The equation $Dh(x) \equiv Dh^\circ(x)$ implies that $h^\circ(x) = h(x) - 2v$, where v is a constant vector. This combined with the condition $|h^\circ(x)| \equiv |h(x)|$ yields that $h(x) - v$ is orthogonal to v. If v was not zero the image of Ω under h would be an $(n-1)$-hyperplane and, consequently, $J(x,h) \equiv 0$. This is impossible for any solution to the equation (13.13), because

$$J(x,h) = \sqrt{\det \mathbf{G}(x, \, |h|)} \neq 0 \tag{13.24}$$

2. Proof of Theorem 13.1

We may assume that the extremal map $h : \mathbb{A} \to \overline{\mathbb{A}^*}$ preserves both the orientation and the order of the boundary components of the annuli. For, if not, we compose h with a suitable conformal automorphism of $\mathbb{A}$. By virtue of Theorem 1.7 any extremal map h is monotone and

$$\mathbb{A}\,(r_*, \, R_*) = \mathbb{A}^* \subset h(\mathbb{A}) \subset \overline{\mathbb{A}^*} = \mathbb{A}\,[r_*, \, R_*] \tag{13.25}$$

Thus, in particular, the set $h^{-1}(\mathbb{A}^*)$ is a connected subset of $\mathbb{A}$. We also consider the annulus $\mathbb{A}(\rho, R) \stackrel{\text{def}}{=\!=} (h^\circ)^{-1}(\mathbb{A}^*) \subset \mathbb{A}$. In case $\operatorname{Mod} \mathbb{A}^* \geqslant \mathcal{N}_\dagger(\operatorname{Mod}\mathbb{A})$ the annulus $\mathbb{A}(\rho, R)$ is the entire annulus $\mathbb{A} = \mathbb{A}(r, R)$. However, when $\operatorname{Mod}\mathbb{A}^* < \mathcal{N}_\dagger(\operatorname{Mod}\mathbb{A})$ the inner radius of the preimage $(h^\circ)^{-1}(\mathbb{A}^*) = \mathbb{A}(\rho, R)$ is determined in such a way that

$$|h^\circ(x)| = r_* \qquad \text{if } r < |x| \leqslant \rho$$

and

$$r_* < |h^\circ(x)| < R_* \qquad \text{if } \rho < |x| < R$$

We shall soon see that $h^{-1}(\mathbb{A}^*) = \mathbb{A}(\rho, R)$ for every extremal map h. To this effect let us consider the union

$$\Omega = h^{-1}(\mathbb{A}^*) \cup \mathbb{A}(\rho, R) \subset \mathbb{A}$$

which is connected because $h^{-1}(\mathbb{A}^*) \cap \mathbb{A}(\rho, R) \neq \emptyset$. This latter set actually contains all points near the outer boundary component of $\mathbb{A}$. Precisely we have $\lim\limits_{|x| \to R} |h(x)| = R_*$. We now make use of the free Lagrangian identities (7.4)

$$\int_{\mathbb{A}} \frac{J(x,h)\,\mathrm{d}x}{|h(x)|^n} = \operatorname{Mod}\mathbb{A}^* = \int_{\mathbb{A}} \frac{J(x,h^\circ)\,\mathrm{d}x}{|h^\circ(x)|^n} \tag{13.26}$$

It follows from (13.12) that

$$\int_\Omega \frac{\eta\big(|h(x)|\big)\,\mathrm{d}x}{|x|^n} = \int_\Omega \frac{\eta\big(|h^\circ(x)|\big)\,\mathrm{d}x}{|x|^n} \tag{13.27}$$

The point to make here is that both Jacobians $J(x,h^\circ)$ and $J(x,h)$ vanish almost everywhere outside Ω. Now, by mean value property for integrals there exists a point $a \in \Omega$ such that

$$\eta\big(|h(a)|\big) = \eta\big(|h^\circ(a)|\big) \tag{13.28}$$

Let $\mathbb{U} \subset \Omega$ denote the set of all points $a \in \Omega$ for which (13.28) holds; $\mathbb{U}$ is certainly relatively closed. This set $\mathbb{U}$ is also open. To see this we first observe that $\eta\big(|h(a)|\big) > 0$ for all $a \in \mathbb{U}$. Indeed, if $a \in \mathbb{A}(\rho, R)$ then $|h^\circ(a)| > r_*$, whereas for $a \in h^{-1}(\mathbb{A}^*)$ we have $|h(a)| > r_*$. It remains to recall that $\eta(\tau) > 0$, whenever $\tau > r_*$. In other words, the Cauchy-Green tensor of h is positive definite near every point of $\mathbb{U}$. Now it is legitimate to appeal to Lemma 13.2. Accordingly, there exists a linear isometry $T : \mathbb{R}^n \to \mathbb{R}^n$ so that $h(x) = T h^\circ(x)$ near the point a; thus, $|h(x)| = |h^\circ(x)|$ near this point. This shows that $\mathbb{U}$ is the entire domain Ω. We then have

$$|h(x)| = |h^\circ(x)| > r_* \qquad \text{for all } x \in \Omega$$

Again by Lemma 13.2 we infer that, upon suitable adjustment via isometry of the target annulus it holds

$$h(x) = h^\circ(x) \qquad \text{in } \Omega$$

As a matter of fact we have

$$h^{-1}(\mathbb{A}^*) = \mathbb{A}(\rho, R)$$

Indeed

$$x \in h^{-1}(\mathbb{A}^*) \;\Leftrightarrow\; h(x) \in \mathbb{A}^* \;\Leftrightarrow\; h^\circ(x) \in \mathbb{A}^* \;\Leftrightarrow\; x \in \mathbb{A}(\rho, R)$$

Thus the uniqueness theorem is proven if $\mathbb{A}(\rho, R) = \mathbb{A}$. This is the case within the Nitsche bounds. In case below the lower Nitsche bound we look at the remaining region $\mathbb{A}(r, \rho] = \mathbb{A} \setminus \Omega$ in which $|h(x)| \equiv r_*$. On the outer boundary of this ring we have $h(x) = h^\circ(x) = r_* \frac{x}{|x|}$. On the other hand, it follows from the equation (13.5) that

$$|h_N| = \frac{\lambda\,|h|}{|x|}\,\eta\big(|h|\big) = \frac{\lambda\, r_*}{|x|}\eta(r_*) = 0 \tag{13.29}$$

Thus h is constant along each ray $t\,\frac{x}{|x|}$, for $r < t \leqslant \rho$. This means that $h(x) = r_*\,\frac{x}{|x|} = h^\circ(x)$ in $\mathbb{A} \setminus \Omega$.

CHAPTER 14

Above the Upper Nitsche Bound, $n \geqslant 4$

This chapter is devoted to a thorough discussion of the minimization problem when, in dimensions $n \geqslant 4$, the target annulus is conformally too fat. We shall see that the extremals are not radially symmetric. But first we need to examine related extremal problems for mappings on spheres.

1. Extremal deformations of the sphere

The study of extremal deformations of ring domains by using spherical coordinates leads us to a variational problem for mappings of the unit sphere. This problem, to be explored later, is the following. Among all homeomorphisms $\Phi : \mathbb{S}^{n-1} \to \mathbb{S}^{n-1}$ of Sobolev class $\mathscr{W}^{1,n}(\mathbb{S}^{n-1}, \mathbb{S}^{n-1})$ find the one which minimizes the energy integral

$$\mathcal{T}[\Phi] = \fint_{\mathbb{S}^{n-1}} \left[\alpha^2 + (n-1)[D\Phi]^2\right]^{\frac{n}{2}} \tag{14.1}$$

where α is any given number. Here $[D\Phi]$ stands for the normalized Hilbert-Schmidt norm of the tangent map

$$D\Phi : T_x\,\mathbb{S}^{n-1} \to T_y\,\mathbb{S}^{n-1}\,, \qquad\qquad y = \Phi(x) \tag{14.2}$$

That is,

$$[D\Phi]^2 = \frac{1}{n-1}\operatorname{Tr}\left[\, D^*\Phi\, D\Phi \,\right] \tag{14.3}$$

which equals 1 for the identity map. An obvious question to ask is whether the identity map

$$id\; :\; \mathbb{S}^{n-1} \to \mathbb{S}^{n-1} \tag{14.4}$$

is the minimizer. Naturally it is tempting to apply Jensen's inequality.

$$\mathcal{T}[\Phi] \geqslant \left[\alpha^2 + (n-1)\fint_{\mathbb{S}^{n-1}} [D\Phi]^2\right]^{\frac{n}{2}} \tag{14.5}$$

In dimensions $n = 2$ and $n = 3$, one may appeal to Hölder's and Hadamard's inequalities to find that

$$\fint_{\mathbb{S}^{n-1}} [D\Phi]^2 \geqslant \left(\fint_{\mathbb{S}^{n-1}} [D\Phi]^{n-1}\right)^{\frac{2}{n-1}} \geqslant \left(\fint_{\mathbb{S}^{n-1}} J(x,\Phi)\, dx\right)^{\frac{2}{n-1}} = 1 \tag{14.6}$$

As usual $J(x, \Phi)$ denotes the Jacobian determinant of $D\Phi$, the pullback via Φ of the standard $(n-1)$-form ω on $\mathbb{S}^{n-1}$.

$$\fint_{\mathbb{S}^{n-1}} J(x,\Phi)\, dx = \deg \Phi = 1 \tag{14.7}$$

Unfortunately, in dimensions greater than 3, the infimum of the integrals

$$\int_{\mathbb{S}^{n-1}} [D\Phi]^2$$

is not attained for the identity map. This infimum equals zero. Here is a computation which, in addition to illustrating this fact, provides a method for constructing more sophisticated examples. Consider a permissible map $\Phi^\epsilon : \mathbb{S}^{n-1} \to \mathbb{S}^{n-1}$, stretching a spherical cap $\mathbb{S}^\epsilon$ of radius ϵ around the north pole onto the entire sphere $\mathbb{S}^{n-1}$.[1] The rest of the sphere is shrank into the south pole.

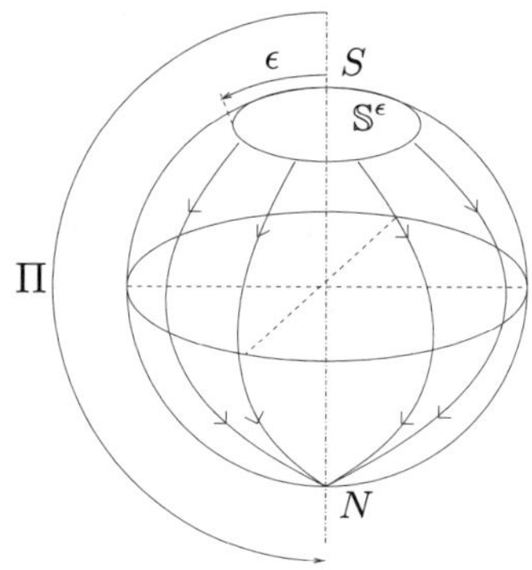

FIGURE 14.1. A spherical cap to be stretched around the sphere.

Elementary geometric considerations show that

$$[D\Phi^\epsilon] \leqslant \begin{cases} \frac{\pi}{\epsilon} & \text{in } \mathbb{S}^\epsilon \\ 0 & \text{otherwise} \end{cases} \tag{14.8}$$

whereas $|\mathbb{S}^\epsilon| \approx \epsilon^{n-1}$. Hence, for every $1 \leqslant p < n-1$ we find that

$$⨍_{\mathbb{S}^{n-1}} [D\Phi^\epsilon(x)]^p\, dx = O(\epsilon^{n-1-p}) \to 0 \qquad \text{as } \epsilon \to 0 \tag{14.9}$$

the infimum being equal to zero. The above computation suggests that we have to express the integrand of $\mathcal{T}[\Phi]$,

$$\left[\alpha^2 + (n-1)[D\Phi]^2\right]^{\frac{n}{2}}$$

as a convex function in $[D\Phi]^{n-1}$. For this purpose we introduce

$$F(s) = \left[\alpha^2 + (n-1)s^{\frac{2}{n-1}}\right]^{\frac{n}{2}}, \qquad 0 \leqslant s < \infty \tag{14.10}$$

Of particular interest to us will be the lower bound of F by a convex function F^* such that

$$\begin{cases} F^* = F^*(s) \leqslant F(s) \\ F^*(1) = F(1) = (\alpha^2 + n - 1)^{\frac{n}{2}} \end{cases} \tag{14.11}$$

---

[1] Φ^ϵ is a weak $\mathscr{W}^{1,n}$-limit of homeomorphisms of $\mathbb{S}^{n-1}$ onto itself.

2. Random variable setting

It is both illuminating and rewarding to consider even more general setting of the variational integrals such as (14.1). We just replace $[D\Phi]^{n-1}$ by a general measurable function.

Let $(\mathbf{S}, \mu)$ be a probability measure space. We shall consider random variables $X : \mathbf{S} \to \mathbb{R}_+$ whose integral mean is at least one;

$$\int_{\mathbf{S}} X \, d\mu \geqslant 1 \tag{14.12}$$

The energy of X is defined by the formula

$$\mathcal{E}[X] = \fint_{\mathbf{S}} \left[\alpha^2 + (n-1) X^{\frac{2}{n-1}}\right]^{\frac{n}{2}} \, d\mu \tag{14.13}$$

where $0 < \alpha < \infty$ and $n = 2, 3, \ldots$ We look for the parameters α for which the constant function $X \equiv 1$ is a minimizer. That is,

$$\inf \mathcal{E}[X] = \mathcal{E}[1] = \left[\alpha^2 + n - 1\right]^{\frac{n}{2}} \tag{14.14}$$

This is certainly true for all α if $n = 2, 3$, because F is convex. However, in higher dimensions F changes concavity. It has an inflection point at

$$s_\circ = \left(\sqrt{\frac{n-3}{n-1}}\,\alpha\right)^{n-1} \tag{14.15}$$

Precisely, F is concave for $0 \leqslant s \leqslant s_\circ$ and convex for $s \geqslant s_\circ$. We shall first examine the case $0 < \alpha \leqslant \alpha_n$, where the upper bound α_n is determined by the equation

$$\left(\alpha_n^2 + n - 1\right)^{\frac{n-2}{2}} \left(\alpha_n^2 - 1\right) = \alpha_n^n \tag{14.16}$$

We have

$$1 < \alpha_n < \sqrt{\frac{n-1}{n-3}} \tag{14.17}$$

Case 1. $0 < \alpha \leqslant \alpha_n$. This means that

$$\left(\alpha^2 + n - 1\right)^{\frac{n-2}{2}} \left(\alpha^2 - 1\right) \leqslant \alpha^n \tag{14.18}$$

It is important to observe that in this case the tangent line of $F = F(s)$ at the point $s = 1$ lies entirely below its graph, see Figure 14.2.

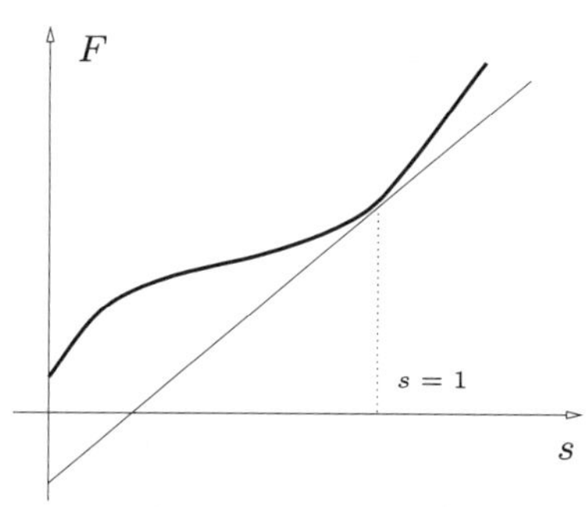

FIGURE 14.2

We then consider a convex lower bound $F^*(X) \leqslant F(X)$ which coincides with that of F for $s \geqslant 1$ and extends along the tangent line for $0 \leqslant s \leqslant 1$. Jensen's inequality yields

$$\begin{aligned} \mathcal{E}[X] &= \int_{\mathbf{S}} F(X)\,d\mu \geqslant \int_{\mathbf{S}} F^*(X)\,d\mu \geqslant F^*\left(\int_{\mathbf{S}} X\,d\mu\right) \\ &\geqslant F^*(1) = F(1) = \mathcal{E}[1] \end{aligned} \tag{14.19}$$

Furhermore, equality occurs if and only if $X \equiv 1$.

Case 2. $\alpha > \alpha_n$. Thus the tangent line to F at $s = 1$ intersect the graph of $F = F(s)$ at some point near the origin, provided α is closed to α_n. This geometric observation suggests that we must look for the extremals which assume exactly two values. The best choice turns out to be when X assumes exactly two values, 0 and $\left(\frac{\alpha}{\alpha_n}\right)^{n-1}$. To see this we split the sample space $\mathbf{S}$ into two parts

$$\mathbf{S} = \mathbf{S}_1 \cup \mathbf{S}_2 \tag{14.20}$$

where

$$\mu(\mathbf{S}_1) = \left(\frac{\alpha_n}{\alpha}\right)^{n-1} \quad \text{and} \quad \mu(\mathbf{S}_2) = 1 - \left(\frac{\alpha_n}{\alpha}\right)^{n-1} \tag{14.21}$$

and define

$$X_\circ = \begin{cases} \left(\frac{\alpha}{\alpha_n}\right)^{n-1} & \text{on } \mathbf{S}_1 \\ 0 & \text{on } \mathbf{S}_2 \end{cases} \tag{14.22}$$

The energy of $X_\circ$ is, therefore, easily computed as

$$\mathcal{E}[X_\circ] = \left[\alpha^2 + (n-1)\frac{\alpha^2}{\alpha_n^2}\right]^{\frac{n}{2}} \mu(\mathbf{S}_1) + \alpha^n\,\mu(\mathbf{S}_2) \tag{14.23}$$

Taking into account the definition of α_n at (14.16), we arrive at the formula

$$\mathcal{E}[X_\circ] = \alpha^n + b\,\alpha \tag{14.24}$$

where

$$b = \frac{n\left(\alpha_n^2 + n - 1\right)^{\frac{n-2}{2}}}{\alpha_n} \tag{14.25}$$

There remains the question as to whether $X_\circ$ possesses the minimum energy among all random variables of mean at least one. To this end we appeal to Lemma 9.1 in its borderline case when $a = a(\alpha_n) = 1$. By formula (9.4) we obtain

$$\left[\alpha^2 + (n-1)X^{\frac{2}{n-1}}\right]^{\frac{n}{2}} \geqslant \alpha^n + b\,\alpha X \tag{14.26}$$

for every random variable $X : \mathbf{S} \to \mathbb{R}_+$. Equality holds if and only if X assumes exactly two values 0 and $\left(\frac{\alpha}{\alpha_n}\right)^{n-1}$. Upon integration over $\mathbf{S}$ we conclude with the desired estimate

$$\mathcal{E}[X_\circ] \geqslant \alpha^n + b\,\alpha \tag{14.27}$$

For the equality to hold the distribution function of the random variable X must coincide with that of $X_\circ$. In particular, the constant function $X \equiv 1$ is not the extremal one.

3. Pulling back a homothety via stereographic projection

Before we see what can happen if $\alpha > \sqrt{\frac{n-1}{n-3}}$, let us consider an example of a non-isometry $\Phi : \mathbb{S}^{n-1} \to \mathbb{S}^{n-1}$ with

$$\fint_{\mathbb{S}^{n-1}} [D\Phi]^{n-1} = 1$$

Here is a construction of such a map.

Let $\Pi : \mathbb{S}^{n-1} \to \widehat{\mathbb{R}}^{n-1}$ denote the stereographic projection of the unit sphere $\mathbb{S}^{n-1} \subset \mathbb{R}^n$ through the south pole onto $\hat{\mathbb{R}}^{n-1}$. Given any positive number λ we consider the homothety $f = f^\lambda : \hat{\mathbb{R}}^{n-1} \to \hat{\mathbb{R}}^{n-1}$ defined by $f^\lambda(x) = \lambda x$ for $x \in \hat{\mathbb{R}}^{n-1}$. Conjugate to f^λ is a conformal mapping $\Phi = \Phi^\lambda : \mathbb{S}^{n-1} \to \mathbb{S}^{n-1}$

$$\Phi^\lambda = \Pi^{-1} \circ f^\lambda \circ \Pi \tag{14.28}$$

We call $\Phi = \Phi^\lambda$ the *spherical homothety*. Let $x = (\cos\theta, \mathfrak{s}\sin\theta) \in \mathbb{S}^{n-1}$ be a point of longitude $\mathfrak{s} \in \mathbb{S}^{n-2} \subset \mathbb{R}^{n-1}$ and meridian $0 \leqslant \theta \leqslant \pi$, see Section 3. The south pole corresponds to $\theta = \pi$. Thus $\Phi : \mathbb{S}^{n-1} \to \mathbb{S}^{n-1}$ is longitude preserving. Let $\varphi = \varphi(\theta)$ denote the meridian of $\Phi(x) = (\cos\varphi, \mathfrak{s}\sin\varphi)$, see Figure 14.3.

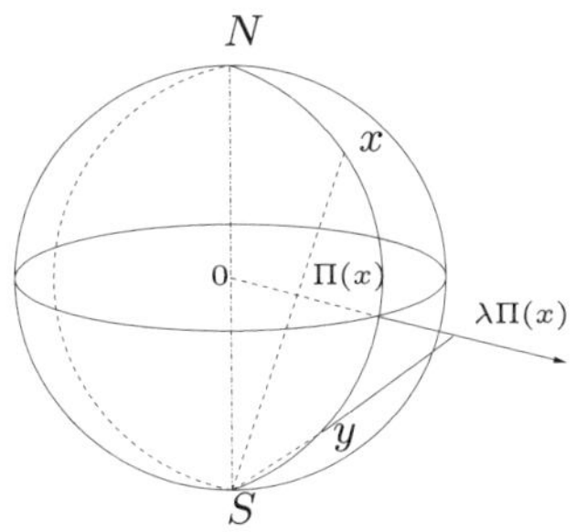

FIGURE 14.3. Spherical homothety $\Phi : \mathbb{S}^{n-1} \to \mathbb{S}^{n-1}$ via stereographic projection.

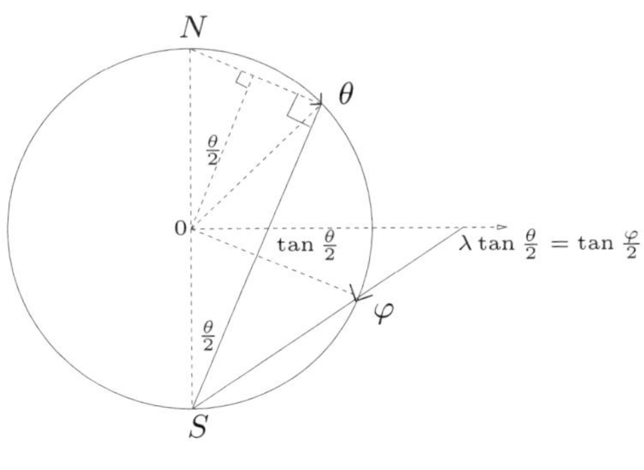

FIGURE 14.4. Computation of the meridian of x and $\Phi(x)$.

Geometric considerations give the following explicit formula

$$\tan\frac{\varphi}{2} = \lambda \tan\frac{\theta}{2} \tag{14.29}$$

see Figure 14.4. Hence

$$\varphi(\theta) = 2\tan^{-1}\left(\lambda \tan\frac{\theta}{2}\right) \tag{14.30}$$

and

$$\begin{gathered} \dot\varphi(\theta) = \frac{1}{\frac{1}{\lambda}\cos^2\frac{\theta}{2} + \lambda\sin^2\frac{\theta}{2}} > 0 \\ \varphi(0) = 0\,,\ \varphi(\pi) = \pi\,,\ \dot\varphi(0) = \lambda\,,\ \dot\varphi(\pi) = \frac{1}{\lambda} \end{gathered} \tag{14.31}$$

see Figure 14.5.

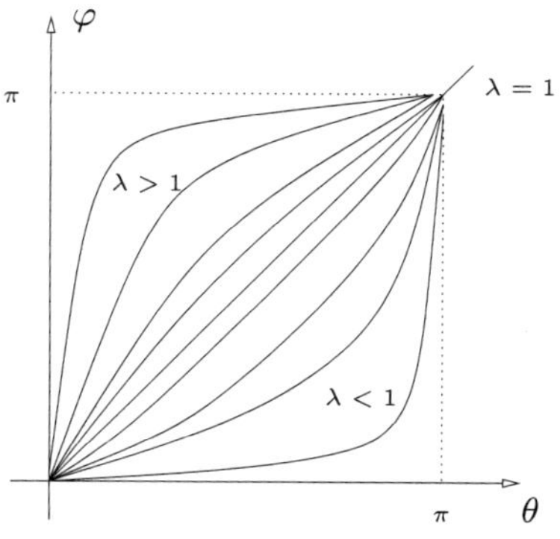

FIGURE 14.5. The meridian functions.

Further differentiation shows that

$$\ddot\varphi(\theta) = \frac{1}{2}\left(\frac{1}{\lambda} - \lambda\right)[\,\dot\varphi(\theta)\,]^2 \sin\theta \tag{14.32}$$

Thus φ is convex if $0 < \lambda < 1$ and concave if $\lambda > 1$. The extreme values of $\dot\varphi$ are assumed at the end-points,

$$\min\left\{\lambda, \frac{1}{\lambda}\right\} \leqslant \dot\varphi(\theta) \leqslant \max\left\{\lambda, \frac{1}{\lambda}\right\} \tag{14.33}$$

Implicit differentiation of (14.29) yields

$$\frac{\dot\varphi}{2\cos^2\frac{\varphi}{2}} = \frac{\lambda}{2\cos^2\frac{\theta}{2}} = \frac{\tan\frac{\varphi}{2}\big/\tan\frac{\theta}{2}}{2\cos^2\frac{\theta}{2}} \tag{14.34}$$

Hence

$$\dot\varphi(\theta) = \frac{\sin\varphi}{\sin\theta} \tag{14.35}$$

for every parameter $\lambda > 0$.

Next, we return to the mapping $\Phi : \mathbb{S}^{n-1} \to \mathbb{S}^{n-1}$ and its tangent bundle map $D\Phi : \mathbf{T}\mathbb{S}^{n-1} \to \mathbf{T}\mathbb{S}^{n-1}$,

$$[D\Phi]^2 \stackrel{\text{def}}{=\!=} \frac{1}{n-1}\left[\dot\varphi^2 + (n-2)\frac{\sin^2\varphi}{\sin^2\theta}\right] = \frac{\sin^2\varphi}{\sin^2\theta} \tag{14.36}$$

Thus

$$[D\Phi] = \frac{\sin\varphi}{\sin\theta} = \dot{\varphi} \tag{14.37}$$

In particular,

$$\min\left\{\lambda, \frac{1}{\lambda}\right\} \leqslant [D\Phi^\lambda] \leqslant \max\left\{\lambda, \frac{1}{\lambda}\right\} \tag{14.38}$$

Finally, let us entertain the reader with the following computation

$$[D\Phi^\lambda]^{n-1} = \det D\Phi^\lambda = \left(\frac{\sin\varphi}{\sin\theta}\right)^{n-1} = \frac{\sin^{n-2}\varphi}{\sin^{n-2}\theta}\,\dot{\varphi} \tag{14.39}$$

Hence we find that

$$\begin{aligned} \fint_{\mathbb{S}^{n-1}} [D\Phi^\lambda]^{n-1} &= \left(\int_0^\pi \dot{\varphi}\,\frac{\sin^{n-2}\varphi}{\sin^{n-2}\theta}\cdot \sin^{n-2}\theta\,d\theta\right)\left(\int_0^\pi \sin^{n-2}\theta\,d\theta\right)^{-1} \\ &= \frac{\int_0^\pi \sin^{n-2}\varphi\,d\varphi}{\int_0^\pi \sin^{n-2}\theta\,d\theta} = 1 \end{aligned} \tag{14.40}$$

as expected.

4. Back to the variational integral $\mathcal{T}[\Phi]$

Recall that

$$\mathcal{T}[\Phi] = \fint_{\mathbb{S}^{n-1}} |\,\alpha^2 + (n-1)[D\Phi]^2\,|^{\frac{n}{2}} \tag{14.41}$$

for homeomorphisms $\Phi : \mathbb{S}^{n-1} \to \mathbb{S}^{n-1}$. We will test it with $\Phi = \Phi^\lambda$, so that

$$[D\Phi]^2 = \frac{1}{n-1}\left[\dot{\varphi}^2 + (n-2)\frac{\sin^2\varphi}{\sin^2\theta}\right] = \frac{\sin^2\varphi}{\sin^2\theta} \tag{14.42}$$

Now we recall the function

$$F(s) = \left[\alpha^2 + (n-1)s^{\frac{2}{n-1}}\right]^{\frac{n}{2}}, \quad s > 0 \tag{14.43}$$

The first and second derivatives of F are given by the formulas

$$\dot{F}(s) = n\left[\alpha^2 + (n-1)s^{\frac{2}{n-1}}\right]^{\frac{n-1}{2}} s^{\frac{3-n}{n-1}} > 0 \tag{14.44}$$

and

$$\ddot{F}(s) = n\left[\alpha^2 + (n-1)s^{\frac{2}{n-1}}\right]^{\frac{n-4}{2}} s^{\frac{4-n}{n-1}}\left(s^{\frac{2}{n-2}} - \frac{n-3}{n-1}\alpha^2\right) \tag{14.45}$$

Hence F is convex if $n = 2, 3$. For $n \geqslant 4$ the function F is concave in the interval

$$0 \leqslant s \leqslant \left(\frac{n-3}{n-1}\right)^{\frac{n-1}{2}} \alpha^{n-1} \tag{14.46}$$

From now on, we consider the case

$$\alpha^2 > \frac{n-1}{n-3}, \qquad n \geqslant 4 \tag{14.47}$$

Given any parameter

$$1 < \lambda \leqslant \sqrt{\frac{n-3}{n-1}}\,\alpha \tag{14.48}$$

We examine the spherical homothety $\Phi = \Phi^\lambda : \mathbb{S}^{n-1} \to \mathbb{S}^{n-1}$. First notice that

$$[D\Phi]^{n-1} \leqslant \left(\frac{n-3}{n-1}\,\alpha^2\right)^{\frac{n-1}{2}} \tag{14.49}$$

point-wise everywhere. Since $[D\Phi]$ is not identically equal to one, by concavity argument we conclude with the following strict inequality

$$\begin{aligned} \fint_{\mathbb{S}^{n-1}} \left[\alpha^2 + (n-1)[D\Phi]^2\right]^{\frac{n}{2}} &= \fint_{\mathbb{S}^{n-1}} \left[\alpha^2 + (n-1)\left([D\Phi]^{n-1}\right)^{\frac{2}{n-1}}\right]^{\frac{n}{2}} \\ &< \left[\alpha^2 + (n-1)\left(\fint_{\mathbb{S}^{n-1}} [D\Phi]^{n-1}\right)^{\frac{2}{n-1}}\right]^{\frac{n}{2}} \\ &= \left[\alpha^2 + n - 1\right]^{\frac{n}{2}} \end{aligned} \tag{14.50}$$

Thus the identity $id : \mathbb{S}^{n-1} \to \mathbb{S}^{n-1}$ is not a minimizer.

5. The failure of radial symmetry, Proof of Theorem 1.11

Throughout this section we make the following standing assumption on the moduli of $\mathbb{A}$ and $\mathbb{A}^*$

$$\operatorname{Mod} \mathbb{A} \leqslant \operatorname{Mod} \mathbb{A}^* \tag{14.51}$$

Theorems 1.9 and 1.14 tell us that in dimensions $n = 2, 3$ the radial mappings are unique (up to a conformal automorphism of $\mathbb{A}$) minimizers of both energy

$$\mathscr{E}_h = \int_{\mathbb{A}} \| Dh \|^n, \qquad h \in \mathscr{P}(\mathbb{A}, \mathbb{A}^*) \tag{14.52}$$

$$\mathscr{F}_h = \int_{\mathbb{A}} \frac{\| Dh \|^n}{|h|^n}, \qquad h \in \mathscr{P}(\mathbb{A}, \mathbb{A}^*) \tag{14.53}$$

This is also true for $n \geqslant 4$ provided the target annulus $\mathbb{A}^*$ is not too fat, as in Theorem 1.10 and Theorem 1.15. Here we construct examples to show that these results do not extend to the full range of moduli at (14.51).

THEOREM 14.1. *Suppose*

$$\operatorname{Mod} \mathbb{A}^* > \sqrt{\frac{n-1}{n-3}} \operatorname{Mod} \mathbb{A}, \qquad n \geqslant 4 \tag{14.54}$$

Then

$$\inf_{h \in \mathscr{R}(\mathbb{A}, \mathbb{A}^*)} \int_{\mathbb{A}} \frac{\| Dh \|^n}{|h|^n} > \inf_{h \in \mathscr{P}(\mathbb{A}, \mathbb{A}^*)} \int_{\mathbb{A}} \frac{\| Dh \|^n}{|h|^n} \tag{14.55}$$

where $\mathscr{R}(\mathbb{A}, \mathbb{A}^*)$ stands for the class of radial stretchings $h : \mathbb{A} \xrightarrow{\text{onto}} \mathbb{A}^*$.

PROOF. In Section 2 we have shown that the infimum in the left hand side is attained for

$$h = h^\alpha(x) = r_* r^{-\alpha} |x|^{\alpha - 1} x, \qquad \alpha = \frac{\operatorname{Mod}\mathbb{A}^*}{\operatorname{Mod}\mathbb{A}} \tag{14.56}$$

The $\mathscr{F}$-energy of h^α is easily computed

$$\mathscr{F}_{h^\alpha} = \left(\alpha^2 + n - 1\right)^{\frac{n}{2}} \operatorname{Mod}\mathbb{A} \tag{14.57}$$

We now test the infimum on the right hand side of (14.55) with the spherical mapping (which is not radial),

$$h(x) = r_* \, r^{-\alpha} |x|^{\alpha-1} \Phi^\lambda \left(\frac{x}{|x|} \right) \tag{14.58}$$

where $\Phi^\lambda : \mathbb{S}^{n-1} \to \mathbb{S}^{n-1}$ is the spherical homothety defined at (14.28). If λ is sufficiently close to 1, but different from 1, then

$$\fint_{\mathbb{S}^{n-1}} \left[\alpha^2 + (n-1)[D\Phi^\lambda]^2 \right]^{\frac{n}{2}} < \left(\alpha^2 + n - 1 \right)^{\frac{n}{2}} \tag{14.59}$$

see (14.50). Now, the $\mathscr{F}$-energy of h is computed as follows

$$\begin{aligned} \mathscr{F}_h &= \int_{\mathbb{A}} \left(\frac{|h_N|^2}{|h|^2} + (n-1) \frac{|h_T|^2}{|h|^2} \right)^{\frac{n}{2}} \\ &= \int_{\mathbb{A}} \left(\alpha^2 + (n-1)[D\Phi^\lambda]^2 \right)^{\frac{n}{2}} \frac{dx}{|x|^n} \\ &= \fint_{\mathbb{S}^{n-1}} \left(\alpha^2 + (n-1)[D\Phi^\lambda]^2 \right)^{\frac{n}{2}} \cdot \operatorname{Mod} \mathbb{A} \\ &< \left(\alpha^2 + n - 1 \right)^{\frac{n}{2}} \operatorname{Mod} \mathbb{A} = \mathscr{F}_{h^\alpha} \end{aligned} \tag{14.60}$$

as desired.

The spherical homothety $\Phi^\lambda : \mathbb{S}^{n-1} \to \mathbb{S}^{n-1}$ works as well for the n-harmonic energy, though a computation is more involved. It results in the proof of Theorem 1.11.

THEOREM 14.2. *Let $n \geqslant 4$ and*

$$\delta_n = \left(\frac{\sqrt{n-1} + \sqrt{n-3}}{\sqrt{n-1} - \sqrt{n-3}} \right)^{\frac{1}{2}} \exp \left[\frac{n-2}{n\sqrt{n-1}} \tan^{-1} \sqrt{n-3} \right] \geqslant \sqrt{n}. \tag{14.61}$$

Consider the annuli $\mathbb{A} = \mathbb{A}(r,R)$ and $\mathbb{A}^* = \mathbb{A}(r_*, R_*)$, such that

$$1 < \frac{R}{r} < \delta_n \quad \textit{and} \quad \frac{R_*}{r_*} > \frac{H_-(\delta_n)}{H_-\left(\delta_n \frac{r}{R}\right)}. \tag{14.62}$$

Then

$$\inf_{h \in \mathscr{R}(\mathbb{A}\,,\,\mathbb{A}^*)} \int_{\mathbb{A}} \| Dh \|^n > \inf_{h \in \mathscr{P}(\mathbb{A}\,,\,\mathbb{A}^*)} \int_{\mathbb{A}} \| Dh \|^n \tag{14.63}$$

PROOF. First we show that the infimum on the left hand side is attained at the radial n-harmonic map

$$h(x) = H\big(|x|\big) \frac{x}{|x|} \tag{14.64}$$

of the form $H(t) = \lambda H_-(kt)$, for suitable parameters $k > \frac{1}{r}$ and $\lambda > 0$. Indeed, the energy of radial mappings is given by

$$\mathscr{E}_h = \omega_{n-1} \int_r^R \left[t^2 \dot{H}^2 + (n-1) H^2 \right]^{\frac{n}{2}} \frac{dt}{t} \tag{14.65}$$

By standard convexity arguments the infimum is attained and the minimizer satisfies the Lagrange-Euler equation. This equation simply means that h is n-harmonic. Since $\operatorname{Mod} \mathbb{A} \leqslant \operatorname{Mod} \mathbb{A}^*$ we are in a situation in which h lies in the class generated by the principal solution H_-; that is, $H(t) = \lambda H_-(kt)$, as claimed. These parameters are uniquely determined by the size of the annuli $\mathbb{A}$ and $\mathbb{A}^*$. Precisely, we have

$$\lambda H_-(kr) = r_* \quad \text{and} \quad \lambda H_-(kR) = R_* \tag{14.66}$$

We eliminate the factor λ by dividing the above equations, and then find k from the equation

$$\frac{H_-(kR)}{H_-(kr)} = \frac{R_*}{r_*} > \frac{R}{r}. \tag{14.67}$$

To find k we observe that the function $Q(t) \overset{\text{def}}{=\!=} \frac{H_-(tr)}{H_-(tR)}$, $\frac{1}{r} < t < \infty$, is increasing from 0 to $\frac{r}{R}$. Indeed

$$\frac{t\dot{Q}(t)}{Q(t)} = \frac{tr\dot{H}_-(tr)}{H_-(tr)} - \frac{tR\dot{H}_-(tR)}{H_-(tR)} = \eta_-(tr) - \eta_-(tR) > 0$$

because the function $\eta_- = \eta_-(s)$ is decreasing, see Figure 5.8 in Chapter 5. This latter statement is immediate from equation (5.29); that is,

$$\left(1 + \frac{\eta_-^2}{n-1}\right)^{\frac{n-2}{2}} (\eta_-^2 - 1) = \frac{-c}{|H(s)|^n}\,, \quad \text{where} \quad H(s) = H_-(s)\,, \text{ for simplicity}$$

Here $-c > 0$ and $|H(s)|$ is increasing. Also note that

$$\lim_{t\to\infty} Q(t) \;=\; \frac{r}{R} \cdot \lim_{t\to\infty} \frac{\frac{H(tr)}{tr}}{\frac{H(tR)}{tR}} = \frac{r}{R} \cdot \frac{\Theta}{\Theta} = \frac{r}{R}$$

see (5.59). Now the equation (14.67) has exactly one solution. Our hypothesis at (14.62) is equivalent to

$$Q(k_\circ) = \frac{r_*}{R_*} \leq Q\left(\frac{\delta_n}{R}\right). \tag{14.68}$$

Since Q is increasing we infer that $k_\circ R \leq \delta_n$. Next, since η_- is decreasing then for $r < t < R$, we have

$$\eta_-(k_\circ t) > \eta_-(k_\circ R) \geqslant \eta_-(\delta_n).$$

Now we appeal to formula (14.61) and (5.47). Accordingly,

$$\Gamma_-\left(\frac{n-3}{n-1}\right) = \delta_n.$$

On the other hand, in view of (5.53) and (5.49)

$$\eta_-(\delta_n) = \frac{1}{u(\delta_n)} = \frac{1}{\Gamma^{-1}(\delta_n)} = \sqrt{\frac{n-1}{n-3}}.$$

Therefore, $\eta_-(k_\circ t) \geqslant \sqrt{\frac{n-1}{n-3}}$ for every $r < t < R$. This means that

$$\eta_{H}(t) > \sqrt{\frac{n-1}{n-3}}\,, \qquad \text{for all } \; r \leqslant t \leqslant R \tag{14.69}$$

Now, we return to the computation of the infimum in the left hand side of (14.63) to obtain

$$
\begin{aligned}
\inf_{h\in\mathscr{R}(\mathbb{A}\,,\,\mathbb{A}^*)} \int_{\mathbb{A}} \| Dh \|^n &= \omega_{n-1} \int_r^R \left[t^2 \dot{H}^2 + (n-1) H^2 \right]^{\frac{n}{2}} \frac{dt}{t} \\
&= \omega_{n-1} \int_r^R [H(t)]^n \left[\eta_{_H}^2(t) + n - 1 \right]^{\frac{n}{2}} \frac{dt}{t}
\end{aligned} \tag{14.70}
$$

Then we test the infimum in the right hand side of (14.63) with the mapping

$$
h_\lambda(x) = H\big(|x|\big)\, \Phi^\lambda \left(\frac{x}{|x|} \right) \tag{14.71}
$$

where, as in the previous case, $\Phi^\lambda : \mathbb{S}^{n-1} \to \mathbb{S}^{n-1}$ is the spherical homothety. We find that

$$
\begin{aligned}
\inf_{h\in\mathscr{P}(\mathbb{A}\,,\,\mathbb{A}^*)} \int_{\mathbb{A}} \| Dh \|^n &\leqslant \int_{\mathbb{A}} \| Dh_\lambda \|^n \\
&= \omega_{n-1} \int_r^R [H(t)]^n \fint_{\mathbb{S}^n} \left[\eta_{_H}^2(t) + (n-1) |D\Phi^\lambda|^2 \right]^{\frac{n}{2}} \frac{dt}{t} \\
&< \omega_{n-1} \int_r^R [H(t)]^n \left[\eta_{_H}^2(t) + n - 1 \right]^{\frac{n}{2}} \frac{dt}{t} \\
&= \inf_{h\in\mathcal{R}(\mathbb{A}\,,\,\mathbb{A}^*)} \int_{\mathbb{A}} \| Dh \|^n
\end{aligned} \tag{14.72}
$$

The above strict inequality follows from (14.50) applied to $\alpha = \eta_{_H}(t) > \sqrt{\frac{n-1}{n-3}}$, where we have chosen $\lambda \neq 1$ sufficiently close to 1.

CHAPTER 15

Quasiconformal Mappings between Annuli

In this final chapter we present an application of free Lagrangians to obtain sharp estimates for quasiconformal homeomorphisms $h : \mathbb{A} \to \mathbb{A}^*$ between annuli $\mathbb{A} = \mathbb{A}(r, R)$ and $\mathbb{A}^* = \mathbb{A}(r_*, R_*)$. With the aid of Möbius transformations (reflections about the spheres and $(n-1)$-dimensional hyperplanes) we may assume that h preserves the orientation and the order of the boundary components. We shall employ the operator norm of the differential matrix, commonly used in the literature [**29, 43, 44**]. Accordingly, a homeomorphism $h : \mathbb{A} \xrightarrow{\text{onto}} \mathbb{A}^*$ of Sobolev class $\mathscr{W}^{1,1}_{\text{loc}}(\mathbb{A}, \mathbb{A}^*)$ is quasiconformal if

$$|Dh(x)|^n \leqslant K\, J(x,h) \qquad \text{a.e. in } \mathbb{A} \tag{15.1}$$

The smallest such number $K \geqslant 1$, denoted by $K_O = K_O(h)$, is called the outer dilatation of h. The inner dilatation is defined to be the smallest number $K_I = K_I(h) \geqslant 1$ such that

$$|D^\sharp h(x)|^n \leqslant K_I\, J(x,h)^{n-1} \qquad \text{a.e. in } \mathbb{A} \tag{15.2}$$

Note the relations

$$K_O \leqslant K_I^{n-1} \quad \text{and} \quad K_I \leqslant K_O^{n-1} \tag{15.3}$$

For the power stretching $h(x) = |x|^{\alpha-1}x$, $\alpha > 0$ we have

$$K_O = \max\left\{\alpha^{-1}\,,\, \alpha^{n-1}\right\} \quad \text{and} \quad K_I = \max\left\{\alpha^{1-n}\,,\, \alpha\right\} \tag{15.4}$$

which shows that both estimates at (15.3) are sharp. Since $J(x,h)$ is integrable and $K < \infty$ we see that $h \in \mathscr{W}^{1,n}(\mathbb{A}, \mathbb{A}^*)$.

THEOREM 15.1. *Suppose $h : \mathbb{A} \xrightarrow{\text{onto}} \mathbb{A}^*$ is a quasiconformal map between annuli. Then*

$$\frac{1}{K_I} \leqslant \left(\frac{\operatorname{Mod}\mathbb{A}^*}{\operatorname{Mod}\mathbb{A}}\right)^{n-1} \leqslant K_O \tag{15.5}$$

This estimate is classic in the theory of quasiconformal mappings, see the pioneering work by F.W. Gehring [**16**]. In the proof below we shall not appeal to any advances in Quasiconformal Theory or PDEs. In fact, our proof provides a method to tackle the uniqueness problem in the borderline cases of (15.5), which seems to be unknown in higher dimensions.

THEOREM 15.2. *If one of the two estimates at (15.5) becomes equality, then it is attained only on the corresponding extremal mappings $h : \mathbb{A} \xrightarrow{\text{onto}} \mathbb{A}^*$ of the form*

$$h(x) = \sqrt{r_*\, R_*}\left(\frac{\sqrt{rR}}{|x|}\right)^{\pm\alpha} \Phi\left(\frac{x}{|x|}\right), \qquad r \leqslant |x| \leqslant R \tag{15.6}$$

where

$$\alpha = \frac{\operatorname{Mod}\mathbb{A}^*}{\operatorname{Mod}\mathbb{A}} = \begin{cases} K_O^{\frac{1}{n-1}} & \text{if } \operatorname{Mod}\mathbb{A}^* \geqslant \operatorname{Mod}\mathbb{A} \\ K_I^{\frac{1}{1-n}} & \text{if } \operatorname{Mod}\mathbb{A}^* \leqslant \operatorname{Mod}\mathbb{A} \end{cases} \tag{15.7}$$

Here the spherical part $\Phi : \mathbb{S}^{n-1} \xrightarrow{onto} \mathbb{S}^{n-1}$ *can be any homeomorphism satisfying*

(i) *volume condition:* $J(\omega, \Phi) \equiv \pm 1$, *for a.e.* $\omega \in \mathbb{S}^{n-1}$
(ii) α*-contraction condition: meaning that*

- $|\, D\Phi(\omega)\,| \leqslant \alpha$, *for the equality in the upper bound of (*15.5*)*
- $|\,(D\Phi(\omega))^{-1}\,| \leqslant 1/\alpha$, *for the equality in the lower bound of (*15.5*)*

Note, that the only volume preserving homeomorphisms $\Phi : \mathbb{S}^1 \xrightarrow{onto} \mathbb{S}^1$ are isometries. Thus, in dimension $n = 2$, the extremal quasiconformal mappings take the form

$$h(z) = \lambda\, |z|^{\alpha-1} z \tag{15.8}$$

where λ is a complex number of modulus $r\, r_*^{-\alpha}$. Example 15.4 shows that in higher dimensions there exist non-isometric homeomorphisms $\Phi : \mathbb{S}^{n-1} \xrightarrow{onto} \mathbb{S}^{n-1}$ which are volume preserving and satisfy the α-contraction condition, see also [**42**]. The exception is the conformal case of $\alpha = 1$ for which the contraction condition $|D\Phi(\omega)| \leqslant 1$ together with $J(\omega, \Phi) \equiv \pm 1$ imply that Φ is an isometry. As a corollary to these observations we obtain Schottky's theorem (1877), [**45**], in $\mathbb{R}^n$.

THEOREM 15.3. *An annulus* $\mathbb{A} = \mathbb{A}(r, R)$ *can be mapped conformally onto* $\mathbb{A}^* = \mathbb{A}(r_*, R_*)$, *if and only if* $\frac{R}{r} = \frac{R_*}{r_*}$. *Moreover, modulo isometry and rescaling, every conformal mapping takes the form*

$$h(x) = \begin{cases} x & \text{the identity} \\ \frac{x}{|x|^2} & \text{the inversion} \end{cases} \tag{15.9}$$

For both Theorems it involves no loss of generality in assuming that h preserves orientation and the order of boundary components of the annuli. And we do so from now on.

0.1. Proof of Theorems 15.1. Let us first prove the inequality at the right hand side of (15.5). We begin with the identity (7.6) and, after Hölder's inequality, use the free Lagrangian (7.4)

$$\begin{aligned} \operatorname{Mod}\mathbb{A}^* &= \int_{\mathbb{A}} \frac{\mathrm{d}|h| \wedge *\mathrm{d}|x|}{|h|\,|x|^{n-1}} = \int_{\mathbb{A}} \left\langle [Dh]^* \frac{h}{|h|}, \frac{x}{|x|} \right\rangle \frac{\mathrm{d}x}{|h|\,|x|^{n-1}} \\ &\leqslant \int_{\mathbb{A}} \frac{|Dh|\,\mathrm{d}x}{|h|\,|x|^{n-1}} \leqslant \left(\int_{\mathbb{A}} \frac{|Dh|^n}{|h|^n} \right)^{\frac{1}{n}} \left(\int_{\mathbb{A}} \frac{\mathrm{d}x}{|x|^n} \right)^{\frac{n-1}{n}} \\ &\leqslant K_O^{\frac{1}{n}} \left(\int_{\mathbb{A}} \frac{J(x,h)\,\mathrm{d}x}{|h(x)|^n} \right)^{\frac{1}{n}} (\operatorname{Mod}\mathbb{A})^{\frac{n-1}{n}} \\ &= K_O^{\frac{1}{n}} (\operatorname{Mod}\mathbb{A}^*)^{\frac{1}{n}} (\operatorname{Mod}\mathbb{A})^{\frac{n-1}{n}} \end{aligned} \tag{15.10}$$

as desired.

For the left hand side at (15.5) we begin with the identity (7.12) to compute in the similar fashion that

$$\begin{aligned}
\operatorname{Mod}\mathbb{A} &= \int_{\mathbb{A}} \frac{\mathrm{d}|x|}{|x|} \wedge h^\sharp\omega = \int_{\mathbb{A}} \left\langle [D^\sharp h]\frac{x}{|x|}, \frac{h}{|h|}\right\rangle \frac{\mathrm{d}x}{|x|\,|h|^{n-1}} \\
&\leqslant \int_{\mathbb{A}} \frac{|D^\sharp h|\,\mathrm{d}x}{|x|\,|h|^{n-1}} \leqslant \left(\int_{\mathbb{A}} \frac{|D^\sharp h|^{\frac{n}{n-1}}}{|h|^n}\right)^{\frac{n-1}{n}} \left(\int_{\mathbb{A}} \frac{\mathrm{d}x}{|x|^n}\right)^{\frac{1}{n}} \\
&\leqslant K_I^{\frac{1}{n}} \left(\int_{\mathbb{A}} \frac{J(x,h)\,\mathrm{d}x}{|h|^n}\right)^{\frac{n-1}{n}} (\operatorname{Mod}\mathbb{A})^{\frac{1}{n}} \qquad (15.11)\\
&= K_I^{\frac{1}{n}} \,(\operatorname{Mod}\mathbb{A}^*)^{\frac{n-1}{n}}\, (\operatorname{Mod}\mathbb{A})^{\frac{1}{n}}
\end{aligned}$$

as desired.

0.2. Proof of Theorem 15.2. Concerning the borderline cases, let $h : \mathbb{A} \overset{\text{onto}}{\longrightarrow} \mathbb{A}^*$ denote the extremal mapping for the right hand side of (15.5). We simplify the matters by assuming that $r = r_* = 1$. Note that the power mapping

$$h^\alpha(x) = |x|^{\alpha-1}x\,, \qquad \alpha = \frac{\operatorname{Mod}\mathbb{A}^*}{\operatorname{Mod}\mathbb{A}} \tag{15.12}$$

is among the extremals. The borderline equation

$$(\operatorname{Mod}\mathbb{A}^*)^{n-1} = K_O\,(\operatorname{Mod}\mathbb{A})^{n-1} \tag{15.13}$$

implies that we have equalities everywhere in (15.10). This amounts to the following two conditions.

(i) $\left\langle D^*h \cdot \frac{h}{|h|}\,, \frac{x}{|x|}\right\rangle \equiv \left|D^*h\cdot \frac{h}{|h|}\right| \equiv |Dh| \equiv \alpha\,\frac{|h|}{|x|}$, α - a positive constant

and

(ii) $|Dh|^n \equiv K_O\,J(x,h)$

The first set of equations is fulfilled if and only if

$$D^*h\cdot \frac{h}{|h|} \equiv |Dh(x)|\,\frac{x}{|x|} \equiv \alpha\frac{|h|}{|x|}\frac{x}{|x|}\,, \qquad \alpha \text{ - a positive constant} \tag{15.14}$$

We identify α from the following equations

$$\frac{\alpha^n}{|x|^n} \equiv \frac{|Dh|^n}{|h|^n} \equiv K_O\frac{J(x,h)}{|h|^n} \tag{15.15}$$

Upon integration over the annulus $\mathbb{A}$ we obtain

$$\alpha^n \operatorname{Mod}\mathbb{A} = K_O \operatorname{Mod}\mathbb{A}^* \tag{15.16}$$

which in view of (15.13) yields

$$\alpha = \frac{\operatorname{Mod}\mathbb{A}}{\operatorname{Mod}\mathbb{A}} = \sqrt[n-1]{K_O} \geqslant 1 \tag{15.17}$$

Now, an interesting nonlinear PDE arises from (i) and, in a more direct manner, from (15.14)

$$D^*h\cdot \frac{h}{|h|^2} = \alpha\frac{x}{|x|^2} \tag{15.18}$$

Rather unexpectedly we can easily solve this equation for $|h|$. Let us express it as

$$\nabla \log|h|^2 \; = \alpha\,\nabla\log|x|^2 \tag{15.19}$$

Hence, in view of the normalization $|h(x)| = 1$ for $|x| = 1$, we find that

$$|h(x)| = |x|^\alpha \tag{15.20}$$

Now, h takes the form

$$h(x) = |x|^\alpha \Psi(x) , \qquad \text{with } \Psi : \mathbb{A} \to \mathbb{S}^{n-1}$$

The map h, being a homeomorphism in $\mathscr{W}^{1,n}(\mathbb{A}, \mathbb{A}^*)$, is differentiable almost everywhere, then so is Ψ. From (15.14) we also find the operator norm of the differential

$$|Dh(x)| = \alpha \frac{|h|}{|x|} = \alpha |x|^{\alpha-1} = |Dh^\alpha(x)| \tag{15.21}$$

This, in view of (ii), yields

$$J(x,h) = \alpha |x|^{n\alpha - n} = J(x, h^\alpha) \tag{15.22}$$

Further examination of the mapping $\Psi : \mathbb{A} \to \mathbb{S}^{n-1}$ will reveal that its normal derivative vanishes almost everywhere. To this end, let $x_\circ \in \mathbb{A}$ be a point of differentiability of Ψ. For small real numbers ϵ we consider Taylor's expansions of order one,

$$\Psi(x_\circ + \epsilon x_\circ) = \Psi(x_\circ) + \epsilon \Psi_N(x_\circ) + o(\epsilon) \tag{15.23}$$

and

$$h(x_\circ + \epsilon x_\circ) - h(x_\circ) = \epsilon \big[Dh(x_\circ)\big] x_\circ + o(\epsilon) \tag{15.24}$$

Upon elementary computation, letting ϵ go to zero, we arrive at the following estimate

$$|\alpha \Psi(x_\circ) + \Psi_N(x_\circ)| \leqslant \alpha \tag{15.25}$$

We square it to obtain

$$\alpha^2 + 2\alpha \langle \Psi \, , \, \Psi_N \rangle + |\Psi_N|^2 \leqslant \alpha^2 \tag{15.26}$$

It is important to observe that Ψ is orthogonal to its directional derivatives, because $\langle \Psi, \Psi \rangle = |\Psi|^2 \equiv 1$. Thus, in particular, $\Psi_N(x_\circ) = 0$. This simply means that

$$\Psi(x) = \Phi\left(\frac{x}{|x|}\right) , \qquad \text{where } \Phi : \mathbb{S}^{n-1} \to \mathbb{S}^{n-1} \tag{15.27}$$

Obviously, $\Phi \in \mathscr{W}^{1,n}(\mathbb{S}^{n-1}, \mathbb{S}^{n-1})$ is a homeomorphism of $\mathbb{S}^{n-1}$ onto itself. It induces the linear tangent map

$$D\Phi(\omega) \, : \, T_\omega \mathbb{S}^{n-1} \to T_\sigma \mathbb{S}^{n-1} \, , \qquad \sigma = \Phi(\omega) \tag{15.28}$$

The Jacobian determinant of Φ, with respect to the standard volume form on $\mathbb{S}^{n-1}$, will be denoted by

$$J(\omega, \Phi) = \det D\Phi \tag{15.29}$$

It relates to the Jacobian determinant of h by the rule

$$J(x,h) = \alpha |x|^{n\alpha - n} \langle \Phi, \Phi_{T_2} \times \ldots \times \Phi_{T_n} \rangle \tag{15.30}$$

$$= \alpha |x|^{n\alpha - n} J(\omega, \Phi) \, , \quad \omega = \frac{x}{|x|} \tag{15.31}$$

Thus, in view of (15.22) we see that $\Phi : \mathbb{S}^{n-1} \to \mathbb{S}^{n-1}$ is volume preserving. Here we can take the cross product of directional derivatives with respect to an arbitrary positively oriented orthonormal frame $T_2, ..., T_n$ in $\mathbf{T}_\omega \mathbb{S}^{n-1}$. Thus, in view of (15.22), we obtain

$$J(\omega, \Phi) = |\Phi_{T_2} \times ... \times \Phi_{T_n}| \leqslant |\Phi_{T_2}| \cdots |\Phi_{T_n}| \leqslant |\, D\Phi \,|^{n-1} \tag{15.32}$$

Next, we look at the Cauchy-Green tensor of h in terms of the orthonormal frame $N, T_2, ..., T_n$ at $T_x\mathbb{A}$

$$\begin{aligned} D^*h\, Dh \;&=\; \begin{bmatrix} - & - & h_N & - & - \\ - & - & h_{T_2} & - & - \\ & & \vdots & & \\ - & - & h_{T_n} & - & - \end{bmatrix} \begin{bmatrix} | & | & | \\ | & | & | \\ h_N & h_{T_2} \cdots & h_{T_n} \\ | & | & | \\ | & | & | \end{bmatrix} \\ &=\; |x|^{2\alpha-2} \begin{bmatrix} - & - & \alpha\Phi & - & - \\ - & - & \Phi_{T_2} & - & - \\ & & \vdots & & \\ - & - & \Phi_{T_n} & - & - \end{bmatrix} \begin{bmatrix} | & | & | \\ | & | & | \\ \alpha\Phi & \Phi_{T_2} \cdots & \Phi_{T_n} \\ | & | & | \\ | & | & | \end{bmatrix} \\ &=\; |x|^{2\alpha-2} \begin{bmatrix} \alpha^2 & 0 & 0 \\ 0 & |\Phi_{T_2}|^2 & \langle \Phi_{T_2}, \Phi_{T_n} \rangle \\ & & \\ 0 & \langle \Phi_{T_n}, \Phi_{T_2} \rangle & |\Phi_{T_n}|^2 \end{bmatrix} \end{aligned} \tag{15.33}$$

and

$$D^*\Phi\, D\Phi = \begin{bmatrix} |\Phi_{T_2}|^2 & \dots & \langle \Phi_{T_2}, \Phi_{T_n} \rangle \\ & \vdots & \\ \langle \Phi_{T_n}, \Phi_{T_2} \rangle & \dots & |\Phi_{T_n}|^2 \end{bmatrix} \tag{15.34}$$

Let $0 \leqslant \lambda_2(x) \leqslant ... \leqslant \lambda_n(x)$ denote the singular values of $D\Phi$, meaning that $\lambda_2^2(x) \leqslant ... \leqslant \lambda_n^2(x)$ are eigenvalues of $D^*\Phi\, D\Phi$. It follows from (15.10) that the numbers

$$\lambda_2(x)|x|^{\alpha-1}, \dots, \lambda_n(x)|x|^{\alpha-1} \text{ and } \alpha|x|^{\alpha-1}$$

are the singular values of Dh. Since $\alpha|x|^{\alpha-1} = |Dh|$, this latter number is the largest singular value,

$$0 \leqslant \lambda_2(x) \leqslant ... \leqslant \lambda_n(x) \leqslant \alpha \tag{15.35}$$

In particular,

$$[D\Phi(x)] \equiv \lambda_n(x) \leqslant \alpha \tag{15.36}$$

as desired.

This computation also shows that the inequality $|D\Phi(x)| \leqslant \alpha$ is both sufficient and necessary for the equation $|Dh(x)| = \alpha|x|^{\alpha-1}$.

A backwards inspection of the above arguments reveals that $h(x) = |x|^\alpha \Phi\left(\frac{x}{|x|}\right)$ is an extremal map if and only if $|D\Phi(x)| \leqslant \alpha$ and $J(\omega, \Phi) \equiv 1$.

It is rewarding to look at the inverse map $f = h^{-1} : \mathbb{A}^* \xrightarrow{\text{onto}} \mathbb{A}$, which takes the form

$$f(y) = |y|^{\frac{1}{\alpha}} \Phi^{-1}\left(\frac{y}{|y|}\right) \tag{15.37}$$

We observe that for every quasiconformal map h and its inverse $h^{-1} = f$ it holds

$$K_O(x,h) = K_I(y,f)\,, \qquad y = h(x) \tag{15.38}$$

Now in much the same way we find all extremals for the left hand side of (15.5). Theorem 15.2 follows.

Let us finish this subsection with an example of the volume preserving mappings of $\mathbb{S}^2 \subset \mathbb{R}^3$.

EXAMPLE 15.4. For every $\epsilon > 0$ there exists a homeomorphism $\Phi \in \mathscr{W}^{1,\infty}(\mathbb{S}^2, \mathbb{S}^2)$, not rotation, such that $|\Phi(\omega) - \omega| \leqslant \epsilon$ and $J(\omega, \Phi) \equiv 1$ for almost every $\omega \in \mathbb{S}^2$.

0.3. Construction of Φ. We shall view Φ as a small perturbation of id : $\mathbb{S}^2 \to \mathbb{S}^2$. The construction of such perturbation is made in three steps. In the first step we project the sphere $x^2 + y^2 + z^2 = 1$ with two poles removed, $z \neq \pm 1$ onto the cylinder $x^2 + y^2 = 1\,,\ -1 < z < 1$. The horizontal rays from the axis of the cylinder project the points of the sphere towards the surface of the cylinder. This projection, sometimes attributed to Archimedes, is well known as Lambert's Cylindrical Projection (Johann H. Lambert 1772). In the second step we cut the cylinder along a path from the top to the bottom circles and unroll it flat. The third map is a piece-wise linear perturbation of the triangle ABC inside the flat region, see Figure 15.1. It keeps A, B, C fixed while permuting X, Y, Z in the cyclic way, $X \to Y \to Z \to X$. It is geometrically clear that such piece-wise linear deformation preserves the area. Moreover, choosing X, Y, Z close to the barycenter makes the deformation arbitrarily close to the identity. Finally we roll it back onto the cylinder and use inverse of Lambert's projection to end up with the desired area preserving perturbation of the identity.

REMARK 15.5. As pointed out by the referee, a simpler example can be given : $(\theta, \varphi) \mapsto (\theta + g(\theta), \varphi)$, where $\theta\,, \varphi$ are the longitude and latitude, and g is any Lipschitz function.

Acknowledgements. We thank Leonid Kovalev for valuable discussions on the subject of this paper and the referee for very careful reading of the paper and many useful suggestions.

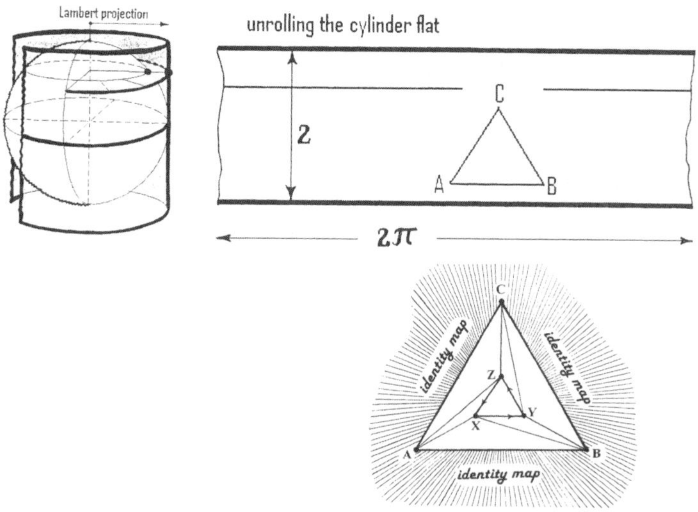

FIGURE 15.1. Constructing an area-preserving Lipschitz homeomorphism.

Bibliography

[1] Antman, S. S. *Fundamental mathematical problems in the theory of nonlinear elasticity.* Univ. Maryland, College Park, Md., (1975), pp. 35–54. MR0468503 (57:8335)

[2] Astala, K., Iwaniec, T., and Martin, G., *Deformations of annuli with smallest mean distortion.* Arch. Ration. Mech. Anal. **195**, no. 3 (2010), 899–921. MR2591976

[3] Astala, K., Iwaniec, T., and Martin, G. J. *Elliptic partial differential equations and quasiconformal mappings in the plane*, Princeton University Press, 2009. MR2472875 (2010j:30040)

[4] Astala, K., Iwaniec, T., Martin, G. J., and Onninen, J. *Extremal mappings of finite distortion.* Proc. London Math. Soc. (3) **91** (2005), no. 3, 655–702. MR2180459 (2006h:30016)

[5] Ball, J. *Convexity conditions and existence theorems in nonlinear elasticity.* Arch. Rational Mech. Anal. **63** (1976/77), no. 4, 337–403. MR0475169 (57:14788)

[6] Ball, J. M., Currie, J. C., and Olver, P. J. *Null Lagrangians, weak continuity, and variational problems of arbitrary order.* J. Funct. Anal. **41** (1981), no. 2, 135–174. MR615159 (83f:49031)

[7] Bojarski, B., and Iwaniec, T. *Analytical foundations of the theory of quasiconformal mappings in R^n.* Ann. Acad. Sci. Fenn. Ser. A I Math. **8** (1983), no. 2, 257–324. MR731786 (85h:30023)

[8] Calabi, E. and Hartman, P. *On the smoothness of isometries.* Duke Math. J. **37** (1970) 741–750. MR0283727 (44:957)

[9] Chen, Y. W. *Discontinuity and representations of minimal surface solutions*, Proceedings of the conference on differential equations (dedicated to A. Weinstein), pp. 115–138. University of Maryland, College Park, MD (1956). MR0086557 (19:203e)

[10] Ciarlet, P. G. *Mathematical elasticity.* Vol. I. Three-dimensional elasticity. Studies in Mathematics and its Applications, 20. North-Holland Publishing Co., Amsterdam (1988). MR936420 (89e:73001)

[11] Cristina, J., Iwaniec, T., Kovalev, L. V., and Onninen, J. *The Hopf-Laplace equation*, arXiv:1011.5934.

[12] Csörnyei, M., Hencl, S., and Malý, J. *Homeomorphisms in the Sobolev space $W^{1,n-1}$*, J. Reine Angew. Math. **644** (2010), 221–235. MR2671780

[13] Duzaar, F., and Mingione, G. *The p-harmonic approximation and the regularity of p-harmonic maps*, Calc. Var. Partial Differential Equations **20** (2004), no. 3, 235–256. MR2062943 (2005d:35094)

[14] Duren, P. *Harmonic mappings in the plane*, Cambridge University Press, Cambridge, 2004. MR2048384 (2005d:31001)

[15] Edelen, D. G. B. *The null set of the Euler-Lagrange operator.* Arch. Rational Mech. Anal. **11** (1962) 117–121. MR0150623 (27:618)

[16] Gehring, F. W. *Rings and quasiconformal mappings in space*, Trans. Amer. Math. Soc. **103** (1962), 353–393. MR0139735 (25:3166)

[17] Han, Z-C. *Remarks on the geometric behavior of harmonic maps between surfaces*, Elliptic and parabolic methods in geometry (Minneapolis, MN, 1994), 57–66. MR1417948 (98a:58048)

[18] Hardt, R. and Lin, F-H. *Mappings minimizing the L^p norm of the gradient*, Comm. Pure Appl. Math. **40** (1987), no. 5, 555–588. MR896767 (88k:58026)

[19] Hencl, S. and Koskela, P. *Regularity of the inverse of a planar Sobolev homeomorphism*, Arch. Ration. Mech. Anal. **180** (2006), no. 1, 75–95. MR2211707 (2007e:30020)

[20] Hencl, S., Koskela, P., and Onninen, J. *A note on extremal mappings of finite distortion*, Math. Res. Lett. **12** (2005), no. 2-3, 231–237. MR2150879 (2006a:30022)

[21] Hildebrandt, S. and Nitsche, J. C. C. *A uniqueness theorem for surfaces of least area with partially free boundaries on obstacles*, Arch. Rational Mech. Anal. **79** (1982), no. 3, 189–218. MR658387 (83k:49068)

[22] Iwaniec, T. *On the concept of the weak Jacobian and Hessian.* Rep. Univ. Jyväskylä Dep. Math. Stat., 83, Univ. Jyväskylä, Jyväskylä, (2001) 181–205. MR1886622 (2003f:30027)

[23] Iwaniec, T., Koskela, P., and Onninen, J. *Mappings of finite distortion: monotonicity and continuity.* Invent. Math. **144** (2001), no. 3, 507–531. MR1833892 (2002c:30029)

[24] Iwaniec, T., Koh, N.-T., Kovalev, L. V., and Onninen, J. *Existence of energy-minimal diffeomorphisms between doubly connected domains*, Invent. Math., to appear.

[25] Iwaniec, T., Kovalev, L. V. and Onninen, J. *Harmonic mappings of an annulus, Nitsche conjecture and its generalizations*, Amer. J. Math. **132** (2010), no. 5, 1397–1428. MR2732352

[26] Iwaniec, T., Kovalev, L. V. and Onninen, J. *The Nitsche conjecture*, J. Amer. Math. Soc. **24** (2011), no. 2, 345–373.

[27] Iwaniec, T., Kovalev, L. V. and Onninen, J. *Harmonic mapping problem and affine capacity*, Proc. Roy. Soc. Edinburgh Sect. A, to appear.

[28] Iwaniec, T., Kovalev, L. V. and Onninen, J. *Doubly connected minimal surfaces and extremal harmonic mappings*, J. Geom. Anal., to appear.

[29] Iwaniec, T. and Martin, G. J. *Geometric Function Theory and Non-linear Analysis.* Oxford Mathematical Monographs. 2001. MR1859913 (2003c:30001)

[30] Iwaniec, T. and Onninen, J. *Deformations of finite conformal energy: boundary behavior and limit theorems*, Trans. Amer. Math. Soc., **363** (2011), no. 11, 5605–5648. MR2817402

[31] Iwaniec, T. and Onninen, J. *Deformations of finite conformal energy: existence and removability of singularities*, Proc. Lond. Math. Soc. (3) **100** (2010), no. 1, 1–23. MR2578466 (2011a:30063)

[32] Iwaniec, T. and Šverák, V. *On mappings with integrable dilatation.* Proc. Amer. Math. Soc. **118** (1993), no. 1, 181–188. MR1160301 (93k:30023)

[33] Jost, J. and Schoen, R. *On the existence of harmonic diffeomorphisms*, Invent. Math., **66**, (1982), 353-359. MR656629 (83i:58033)

[34] Kalaj, D., *On the Nitsche conjecture for harmonic mappings in* $\mathbb{R}^2$ *and* $\mathbb{R}^3$. Israel J. Math. **150** (2005), 241–251. MR2255810 (2007i:31001)

[35] Lyzzaik, A. *The modulus of the image annuli under univalent harmonic mappings and a conjecture of J.C.C. Nitsche*, J. London Math. Soc., **64**, (2001), 369–384. MR1853457 (2002h:30023)

[36] Mingione, G. *Regularity of minima: an invitation to the dark side of the calculus of variations*, Appl. Math. **51** (2006), no. 4, 355–426. MR2291779 (2007k:35086)

[37] Morrey, C.B. *The Topology of (Path) Surfaces.* Amer. J. Math. **57** (1935), no. 1, 17–50. MR1507053

[38] Morrey, C.B. *Quasi-convexity and the lower semicontinuity of multiple integrals.* Pacific J. Math. **2**, (1952) 25–53. MR0054865 (14:992a)

[39] Morrey, C.B. *A variational method in the theory of harmonic integrals. II*, Amer. J. Math. **78** (1956), 137–170. MR0087765 (19:408a)

[40] Morrey, C.B. *Multiple integrals in the calculus of variations*, Springer-Verlag (1966). MR0202511 (34:2380)

[41] Nitsche, J.C.C. *On the modulus of doubly connected regions under harmonic mappings*, Amer. Math. Monthly, **69**, (1962), 781–782. MR1531840

[42] Rajala, K. *Distortion of quasiconformal and quasiregular mappings at extremal points.* Michigan Math. J. **53** (2005), no. 2, 447–458. MR2152710 (2006b:30046)

[43] Reshetnyak, Yu. G. *Space mappings with bounded distortion.* American Mathematical Society, Providence, RI, 1989. MR994644 (90d:30067)

[44] Rickman, S. *Quasiregular mappings.* Springer-Verlag, Berlin, 1993. MR1238941 (95g:30026)

[45] Schottky, F.H. *Über konforme Abbildung von mehrfach zusammenhängenden Fläche.* J. für Math., **83**, (1877).

[46] Turowski, G. *Behaviour of doubly connected minimal surfaces at the edges of the support surface*, Arch. Math. (Basel) **77** (2001), no. 3, 278–288. MR1865870 (2002h:53016)

[47] Uhlenbeck, K. *Regularity for a class of non-linear elliptic systems*, Acta Math. **138** (1977), no. 3-4, 219–240. MR0474389 (57:14031)

[48] Uraltseva, N. *Degenerate quasilinear elliptic systems*, Zapiski Nauch. Sem. LOMI, **7**, (1968), 184–222. MR0244628 (39:5942)

[49] Wang, C. *A compactness theorem of n-harmonic maps*, Ann. Inst. H. Poincaré Anal. Non Linéaire **22** (2005), no. 4, 509–519. MR2145723 (2006a:58022)

[50] Weitsman, A. *Univalent harmonic mappings of annuli and a conjecture of J.C.C. Nitsche*, Israel J. Math., **124**, (2001), 327–331. MR1856525 (2002f:30011)

Editorial Information

To be published in the *Memoirs*, a paper must be correct, new, nontrivial, and significant. Further, it must be well written and of interest to a substantial number of mathematicians. Piecemeal results, such as an inconclusive step toward an unproved major theorem or a minor variation on a known result, are in general not acceptable for publication.

Papers appearing in *Memoirs* are generally at least 80 and not more than 200 published pages in length. Papers less than 80 or more than 200 published pages require the approval of the Managing Editor of the Transactions/Memoirs Editorial Board. Published pages are the same size as those generated in the style files provided for $\mathcal{A}\mathcal{M}\mathcal{S}$-LaTeX or $\mathcal{A}\mathcal{M}\mathcal{S}$-TeX.

Information on the backlog for this journal can be found on the AMS website starting from `http://www.ams.org/memo`.

A Consent to Publish and Copyright Agreement is required before a paper will be published in the *Memoirs*. After a paper is accepted for publication, the Providence office will send a Consent to Publish and Copyright Agreement to all authors of the paper. By submitting a paper to the *Memoirs*, authors certify that the results have not been submitted to nor are they under consideration for publication by another journal, conference proceedings, or similar publication.

Information for Authors

Memoirs is an author-prepared publication. Once formatted for print and on-line publication, articles will be published as is with the addition of AMS-prepared frontmatter and backmatter. Articles are not copyedited; however, confirmation copy will be sent to the authors.

Initial submission. The AMS uses Centralized Manuscript Processing for initial submissions. Authors should submit a PDF file using the Initial Manuscript Submission form found at `www.ams.org/submission/memo`, or send one copy of the manuscript to the following address: Centralized Manuscript Processing, MEMOIRS OF THE AMS, 201 Charles Street, Providence, RI 02904-2294 USA. If a paper copy is being forwarded to the AMS, indicate that it is for *Memoirs* and include the name of the corresponding author, contact information such as email address or mailing address, and the name of an appropriate Editor to review the paper (see the list of Editors below).

The paper must contain a *descriptive title* and an *abstract* that summarizes the article in language suitable for workers in the general field (algebra, analysis, etc.). The *descriptive title* should be short, but informative; useless or vague phrases such as "some remarks about" or "concerning" should be avoided. The *abstract* should be at least one complete sentence, and at most 300 words. Included with the footnotes to the paper should be the 2010 *Mathematics Subject Classification* representing the primary and secondary subjects of the article. The classifications are accessible from `www.ams.org/msc/`. The Mathematics Subject Classification footnote may be followed by a list of *key words and phrases* describing the subject matter of the article and taken from it. Journal abbreviations used in bibliographies are listed in the latest *Mathematical Reviews* annual index. The series abbreviations are also accessible from `www.ams.org/msnhtml/serials.pdf`. To help in preparing and verifying references, the AMS offers MR Lookup, a Reference Tool for Linking, at `www.ams.org/mrlookup/`.

Electronically prepared manuscripts. The AMS encourages electronically prepared manuscripts, with a strong preference for $\mathcal{A}\mathcal{M}\mathcal{S}$-LaTeX. To this end, the Society has prepared $\mathcal{A}\mathcal{M}\mathcal{S}$-LaTeX author packages for each AMS publication. Author packages include instructions for preparing electronic manuscripts, samples, and a style file that generates the particular design specifications of that publication series. Though $\mathcal{A}\mathcal{M}\mathcal{S}$-LaTeX is the highly preferred format of TeX, author packages are also available in $\mathcal{A}\mathcal{M}\mathcal{S}$-TeX.

Authors may retrieve an author package for *Memoirs of the AMS* from `www.ams.org/journals/memo/memoauthorpac.html` or via FTP to `ftp.ams.org` (login as `anonymous`, enter your complete email address as password, and type `cd pub/author-info`). The

AMS Author Handbook and the *Instruction Manual* are available in PDF format from the author package link. The author package can also be obtained free of charge by sending email to `tech-support@ams.org` or from the Publication Division, American Mathematical Society, 201 Charles St., Providence, RI 02904-2294, USA. When requesting an author package, please specify $\mathcal{A}\mathcal{M}\mathcal{S}$-LaTeX or $\mathcal{A}\mathcal{M}\mathcal{S}$-TeX and the publication in which your paper will appear. Please be sure to include your complete mailing address.

After acceptance. The source files for the final version of the electronic manuscript should be sent to the Providence office immediately after the paper has been accepted for publication. The author should also submit a PDF of the final version of the paper to the editor, who will forward a copy to the Providence office.

Accepted electronically prepared files can be submitted via the web at `www.ams.org/submit-book-journal/`, sent via FTP, or sent on CD to the Electronic Prepress Department, American Mathematical Society, 201 Charles Street, Providence, RI 02904-2294 USA. TeX source files and graphic files can be transferred over the Internet by FTP to the Internet node `ftp.ams.org` (130.44.1.100). When sending a manuscript electronically via CD, please be sure to include a message indicating that the paper is for the *Memoirs*.

Electronic graphics. Comprehensive instructions on preparing graphics are available at `www.ams.org/authors/journals.html`. A few of the major requirements are given here.

Submit files for graphics as EPS (Encapsulated PostScript) files. This includes graphics originated via a graphics application as well as scanned photographs or other computer-generated images. If this is not possible, TIFF files are acceptable as long as they can be opened in Adobe Photoshop or Illustrator.

Authors using graphics packages for the creation of electronic art should also avoid the use of any lines thinner than 0.5 points in width. Many graphics packages allow the user to specify a "hairline" for a very thin line. Hairlines often look acceptable when proofed on a typical laser printer. However, when produced on a high-resolution laser imagesetter, hairlines become nearly invisible and will be lost entirely in the final printing process.

Screens should be set to values between 15% and 85%. Screens which fall outside of this range are too light or too dark to print correctly. Variations of screens within a graphic should be no less than 10%.

Inquiries. Any inquiries concerning a paper that has been accepted for publication should be sent to `memo-query@ams.org` or directly to the Electronic Prepress Department, American Mathematical Society, 201 Charles St., Providence, RI 02904-2294 USA.

Editors

Selected Titles in This Series

1027 **Olivier Druet, Frédéric Robert, and Juncheng Wei,** The Lin-Ni's problem for mean convex domains, 2012

1026 **Mark Behrens,** The Goodwillie Tower and the EHP sequence, 2012

1025 **Joel Smoller and Blake Temple,** General relativistic self-similar waves that induce an anomalous acceleration into the standard model of cosmology, 2012

1024 **Mats Boij, Juan C Migliore, Rosa M. Miró-Roig, Uwe Nagel, and Fabrizio Zanello,** On the shape of a pure O-sequence, 2012

1023 **Tadeusz Iwaniec and Jani Onninen,** n-Harmonic mappings between annuli: The art of integrating free Lagrangians, 2012

1022 **Maurice Duits, Arno B. J. Kuijlaars, and Man Yue Mo,** The Hermitian two matrix model with an even quartic potential, 2012

1021 **Arnaud Deruelle, Katura Miyazaki, and Kimihiko Motegi,** Networking Seifert surgeries on knots, 2012

1020 **Dominic Joyce and Yinan Song,** A theory of generalized Donaldson-Thomas invariants, 2012

1019 **Abdelhamid Meziani,** On first and second order planar elliptic equations with degeneracies, 2012

1018 **Nicola Gigli,** Second order analysis on $(\mathscr{P}_2(M), W_2)$, 2012

1017 **Zenon Jan Jabłoński, Il Bong Jung, and Jan Stochel,** Weighted shifts on directed trees, 2012

1016 **Christophe Breuil and Vytautas Paškūnas,** Towards a modulo p Langlands correspondence for GL_2, 2012

1015 **Jun Kigami,** Resistance forms, quasisymmetric maps and heat kernel estimates, 2012

1014 **R. Fioresi and F. Gavarini,** Chevalley supergroups, 2012

1013 **Kaoru Hiraga and Hiroshi Saito,** On L-packets for inner forms of SL_n, 2012

1012 **Guy David and Tatiana Toro,** Reifenberg parameterizations for sets with holes, 2012

1011 **Nathan Broomhead,** Dimer models and Calabi-Yau algebras, 2012

1010 **Greg Kuperberg and Nik Weaver,** A von Neumann algebra approach to quantum metrics/Quantum relations, 2012

1009 **Tarmo Järvilehto,** Jumping numbers of a simple complete ideal in a two-dimensional regular local ring, 2011

1008 **Lee Mosher, Michah Sagee, and Kevin Whyte,** Quasi-actions on trees II: Finite depth Bass-Serre trees, 2011

1007 **Steve Hofmann, Guozhen Lu, Dorina Mitrea, Marius Mitrea, and Lixin Yan,** Hardy spaces associated to non-negative self-adjoint operators satisfying Davies-Gaffney estimates, 2011

1006 **Theo Bühler,** On the algebraic foundations of bounded cohomology, 2011

1005 **Frank Duzaar, Giuseppe Mingione, and Klaus Steffen,** Parabolic systems with polynomial growth and regularity, 2011

1004 **Michael Handel and Lee Mosher,** Axes in outer space, 2011

1003 **Palle E. T. Jorgensen, Keri A. Kornelson, and Karen L. Shuman,** Iterated function systems, moments, and transformations of infinite matrices, 2011

1002 **Man Chun Leung,** Supported blow-up and prescribed scalar curvature on S^n, 2011

1001 **N. P. Strickland,** Multicurves and equivariant cohomology, 2011

1000 **Toshiyuki Kobayashi and Gen Mano,** The Schrödinger model for the minimal representation of the indefinite orthogonal group $O(p,q)$, 2011

999 **Montserrat Casals-Ruiz and Ilya Kazachkov,** On systems of equations over free partially commutative groups, 2011

998 **Guillaume Duval,** Valuations and differential Galois groups, 2011

997 **Hideki Kosaki,** Positive definiteness of functions with applications to operator norm inequalities, 2011